EDWARD CARPENTER

CIVILISATION

ITS CAUSE AND CURE

AND OTHER ESSAYS

Elibron Classics
www.elibron.com

Elibron Classics series.

ISBN 1-4021-8613-4 (paperback)
ISBN 1-4021-1730-2 (hardcover)

This Elibron Classics Replica Edition is an unabridged facsimile of the edition published in 1906 by Swan Sonnenschein & Co., Lim., London.

CIVILISATION

ITS CAUSE AND CURE

AND OTHER ESSAYS

(New and Enlarged Edition)

BY

EDWARD CARPENTER

AUTHOR OF "ENGLAND'S IDEAL," "TOWARDS DEMOCRACY," "FROM ADAM'S PEAK TO ELEPHANTA," ETC.

LONDON
SWAN SONNENSCHEIN & CO., Lim.
NEW YORK: CHARLES SCRIBNER'S SONS
1906

SOME PRESS OPINIONS

"Mr. Carpenter never wrote anything in all his books so good as 'Civilisation: Its Cause and Cure.' It expresses in a delightful manner the best and most charming of all the doubts of our time. Civilisation, having destroyed so many idols, has at last suddenly been discovered to be itself an idol, and a destructible idol; and the true sceptic, who has so often asked in the name of civilisation whether there was any reason why we should care for creeds or traditions or unaccountable loyalties, has finally and logically come to asking, in the name of common sense, if there is any reason why we should care for civilisation."—*Daily News*.

"A strange but suggestive paper, entitled 'Civilisation: Its Cause and Cure,' takes up the greater part of the volume, and gives it its title. The fact of disease, mental and physical, as characteristic of the stage in the evolution of each race, called civilisation, is dwelt on in the first part of the paper. The cause and purpose in history of this failure of human health is treated of in the second part."—*Ruskin Reading Guild*.

"Edward Carpenter is a seer. So say those who know him best; and the reader of this little volume of essays, in which he has put his ripest thoughts, will be inclined to agree with that judgment. He sees into things farther than other men, and has the faculty of looking through the crust of convention and artificiality in which the modern world is embedded. The title of the first essay, which also entitles the book, is a bold one, and suggests the author's aim. He holds that the rather indefinite thing which we mean by civilisation has, if used in the definite sense adopted by writers on primitive society, obscured the real life of man, and brought about an unhealthy change in the structure of human society. In its historic evolution he finds that these bad changes have been coincident with the growth of property and the ideas flowing from it. And he goes on to argue that the social life of the wilder races is more harmonious and compact than that of civilised nations. It may sound startling and paradoxical to say this, but the reader who dismisses such theories as absurd will regret that he has not followed the argument through."—*Bradfield Observer*

CONTENTS.

First Edition, *June 1889*; Second Edition, *December 1890*;
Third Edition, *November 1893*; Fourth Edition, *July 1895*;
Fifth Edition, *September 1897*; Sixth Edition, *October 1900*;
Seventh Edition, *July 1902*; Eighth Edition, *March 1903*;
Ninth Edition, *March 1906*.

CIVILISATION: ITS CAUSE AND CURE.

—:o:—

The friendly and flowing savage, who is he? Is he waiting for civilisation, or is he past it, and mastering it?—WHITMAN.

WE find ourselves to-day in the midst of a somewhat peculiar state of society, which we call Civilisation, but which even to the most optimistic among us does not seem altogether desirable. Some of us, indeed, are inclined to think that it is a kind of disease which the various races of man have to pass through—as children pass through measles or whooping cough; but if it is a disease, there is this serious consideration to be made, that while History tells us of many nations that have been attacked by it, of many that have succumbed to it, and of some that are still in the throes of it, we know of no single case in which a nation has fairly recovered from and passed through it to a more normal and healthy condition. In other words the development of human society has never yet (that we know of) passed beyond a certain definite and apparently final stage in the process we call Civilisation; at that stage it has always succumbed or been arrested.

Of course it may at first sound extravagant to use the word disease in connection with Civilisation at all, but a little thought should show that the association is not ill-grounded. To take the matter on its physical side first, I find that in Mullhall's Dictionary of Statistics the number of accredited doctors and surgeons in the United Kingdom is put at over 23,000. If the extent of the national sickness is such that we require 23,000 medical men to attend to us, it must surely be rather serious! And *they* do not cure us. Wherever we look to-day, in mansion or in slum, we see the features and hear the complaints of ill-health; the difficulty is really to find a healthy person. The state of the modern civilised man in this respect—our coughs, colds, mufflers, dread of a waft of chill air, &c.—is anything but creditable, and it seems to be the fact that, notwithstanding all our libraries of medical science, our knowledges, arts, and appliances of life, we are actually less capable of taking care of ourselves than the animals are. Indeed, talking of animals, we are—as Shelley I think points out—fast depraving the *domestic* breeds. The cow, the horse, the sheep, and even the confiding pussy-cat, are becoming ever more and more subject to disease, and are liable to ills which in their wilder state they knew not of. And finally the savage races of the earth do not escape the baneful influence. Wherever Civilisation touches them, they die like flies from the small-pox, drink, and worse evils it brings along with it; and often its mere contact is sufficient to destroy whole races.

But the word Disease is applicable to our social as well as to our physical condition. For as in the body disease arises from the loss of the physical unity which constitutes Health, and so takes the form of warfare or discord between the various parts, or of the abnormal development of individual organs, or the consumption of the system by predatory germs and growths; so in our modern life we find the unity gone which constitutes true society, and in its place warfare of classes and individuals, abnormal development of some to the detriment of others, and consumption of the organism by masses of social parasites. If the word disease is applicable anywhere, I should say it is—both in its direct and its derived sense—to the civilised societies of to-day.

Again, mentally, is not our condition anything but satisfactory? I am not alluding to the number and importance of the lunatic asylums which cover our land, nor to the fact that maladies of the brain and nervous system are now so common; but to the strange sense of mental unrest which marks our populations, and which amply justifies Ruskin's cutting epigram: that our two objects in life are, "Whatever we have—to get more; and wherever we are—to go somewhere else." This sense of unrest, of disease, penetrates down even into the deepest regions of man's being—into his moral nature—disclosing itself there, as it has done in all nations notably at the time of their full civilisation, as the sense of Sin. All down the Christian centuries we find this strange sense of inward strife and discord developed, in marked contrast to the naive insouciance of the pagan and primitive world; and, what is strangest, we even find people glorying in this consciousness—which, while it may be the harbinger of better things to come, is and can be in itself only the evidence of loss of unity and therefore of ill-health, in the very centre of human life.

Of course we are aware with regard to Civilisation that the word is sometimes used in a kind of ideal sense, as to indicate a state of future culture towards which we are tending—the implied assumption being that a sufficiently long course of top hats and telephones will in the end bring us to this ideal condition; while any little drawbacks in the process, such as we have just pointed out, are explained as being merely accidental and temporary. Men sometimes speak of civilising and ennobling influences as if the two terms were interchangeable, and of course if they like to use the word Civilisation in this sense they have a right to; but whether the actual tendencies of modern life taken in the mass *are* ennobling (except in a quite indirect way hereafter to be dwelt upon) is to say the least a doubtful question. Any one who would get an idea of the glorious being that is as a matter of fact being turned out by the present process should read Mr. Kay Robinson's article in the *Nineteenth Century* for May, 1883, in which he prophesies (quite solemnly and in the name of science) that the human being of the future will be a toothless, bald, toeless creature with flaccid muscles and limbs almost incapable of locomotion!

Perhaps it is safer on the whole not to use the word Civilisation in such ideal sense, but to limit its use (as is done to-day by all writers on primitive society) to a definite historical stage through which the various nations pass, and in which we actually find ourselves at the present time. Though there is of course a difficulty in marking the commencement of any period of historical evolution very definitely, yet all students of this subject agree that the growth of property and the ideas and institutions flowing from it did at a certain point bring about such a change in the structure of human society that the new stage might fairly be distinguished from the earlier stages of Savagery and Barbarism by a separate term. The growth of wealth, it is shown, and with it the conception of private property, brought on certain very definite new forms of social life; it destroyed the ancient system of society based upon the *gens*, that is, a society of equals founded upon blood-relationship, and introduced a society of classes founded upon differences of material possession; it destroyed the ancient system of mother-right and inheritance through the female line, and turned the woman into the property of the man; it brought with it private ownership of land, and so created a class of landless aliens, and a whole system of rent, mortgage, interest, &c.; it introduced slavery, serfdom and wage-labor, which are only various forms of the dominance of one class over another; and to rivet these authorities it created the State and the policeman. Every race that we know, that has become what we call civilised, has passed thro' these changes; and though the details may vary and have varied a little, the main order of change has been practically the same in all cases. We are justified therefore in calling Civilisation a historical stage, whose commencement dates roughly from the division of society into classes founded on property, and the adoption of class-government. Lewis Morgan in his *Ancient Society* adds the invention of writing and the consequent adoption of written History and written Law; Engels in his *Ursprung der Familie, des Privat-eigenthums und des Staats* points out the importance of the appearance of the Merchant, even in his most primitive form, as a mark of the civilisation-period; while the French writers of the last century made a good point in inventing the term *nations policées* (policemanised nations) as a substitute for civilised nations; for perhaps there is no better or more universal mark of the period we are considering, and of its social degradation, than the appearance of the crawling phenomenon in question. [Imagine the rage of any decent North American Indians if they had been told they required *policemen* to keep them in order!]

If we take this historical definition of Civilisation, we shall see that our English Civilisation began hardly more than a thousand years ago, and even so the remains of the more primitive society lasted long after that. In the case of Rome—if we reckon from the later times of the early kings down to the fall of Rome—we have again about a thousand years. The Jewish civilisation from David and Solomon downwards lasted—with breaks—somewhat over a thousand years; the Greek civilisation less; the Egyptian considerably more; but the important points to see are, first, that the process has been quite similar in character in these various (and numerous other) cases,[1] quite as similar in fact as the course of the same disease in various persons; and secondly that in no case, as said before, has any nation come *through* and passed beyond this stage; but that in most cases it has succumbed soon after the main symptoms had been developed.

But it will be said, It may be true that civilisation regarded as a stage of human history presents some features of disease; but is there any reason for supposing that disease in some form or other was any less present in the previous stage—that of Barbarism? To which I reply, I think there is good reason. Without committing ourselves to the unlikely theory that the "noble savage" was an ideal human being physically or in any other respect, and while certain that in many points he was decidedly inferior to the civilised man, I think we must allow him the superiority in some directions; and one of these was his comparative freedom from disease. Lewis Morgan, who grew up among the Iroquois Indians, and who probably knew the North American natives as well as any white man has ever done, says (in his *Ancient Society*, p. 45), "Barbarism ends with the production of grand Barbarians." And though there are no native races on the earth to-day who are actually in the latest and most advanced stage of Barbarism[2]; yet if we take the most advanced tribes that we know of—such as the said Iroquois Indians of twenty or thirty years ago, some of the Kaffir tribes round Lake Nyassa in Africa, now (and possibly for a few years more) comparatively untouched by civilisation,

[1] For proof I must refer the reader to Engels, or to his own studies of history.

[2] Say like the Homeric Greeks, or the Spartans of the Lycurgus period.

or the tribes along the river Uaupes, 30 or 40 years back, of Wallace's *Travels on the Amazon*—all tribes in what Morgan would call the *middle* stage of Barbarism—we undoubtedly in each case discover a fine and (which is our point here) *healthy* people. Captain Cook in his first Voyage says of the natives of Otaheite, "We saw no critical disease during our stay upon the island, and but few instances of sickness, which were accidental fits of the colic;" and, later on, of the New Zealanders, "They enjoy perfect and uninterrupted health. In all our visits to their towns, where young and old, men and women, crowded about us . . . . we never saw a single person who appeared to have any bodily complaint, nor among the numbers we have seen naked did we once perceive the slightest eruption upon the skin, or any marks that an eruption had left behind." These are pretty strong words. Of course diseases exist among such peoples, even where they have never been in contact with civilisation, but I think we may say that among the higher types of savages they are rarer, and nothing like so various and so prevalent as they are in our modern life; while the power of recovery from *wounds* (which are of course the most frequent form of disablement) is generally admitted to be something astonishing. Speaking of the Kaffirs, J. G. Wood says, "Their state of health enables them to survive injuries which would be almost instantly fatal to any civilised European." Mr. Frank Oates in his Diary[1] mentions the case of a man who was condemned to death by the king. He was hacked down with axes, and left for dead. "What must have been intended for the *coup de grâce* was a cut in the back of the head, which had chipped a large piece out of the skull, and must have been meant to cut the spinal cord where it joins the brain. It had however been made a little higher than this, but had left such a wound as I should have

[1] *Matabele Land and the Victoria Falls.* p. 209.

thought that no one could have survived . . . when I held the lanthorn to investigate the wound I started back in amazement to see a hole at the base of the skull, perhaps two inches long and an inch and a half wide, and I will not venture to say how deep, but the depth too must have been an affair of inches. Of course this hole penetrated into the substance of the brain, and probably for some distance. I dare say a mouse could have sat in it." Yet the man was not so much disconcerted. Like Old King Cole, "He asked for a pipe and a drink of brandy," and ultimately made a perfect recovery! Of course it might be said that such a story only proves the lowness of organisation of the brains of savages; but to the Kaffirs at any rate this would not apply; they are a quick-witted race, with large brains, and exceedingly acute in argument, as Colenso found to his cost. Another point which indicates superabundant health is the amazing animal spirits of these native races! The shouting, singing, dancing kept up nights long among the Kaffirs are exhausting merely to witness, while the graver North American Indian exhibits a corresponding power of life in his eagerness for battle or his stoic resistance of pain.[1]

Similarly when we come to consider the social life of the wilder races—however rudimentary and undeveloped it may be—the almost universal testimony of students and travelers is that within its limits it is more harmonious and compact than that of the civilised nations. The members of the tribe are not organically at warfare with each other; society is not divided into classes which prey upon each other; nor is it

[1] A similar physical health and power of life are also developed among Europeans who have lived for long periods in more native conditions. It is not to our *race*, which is probably superior to any in capacity, but to the state in which we live that we must ascribe our defect in this particular matter.

consumed by parasites. There is more true social unity, less of disease. Though the customs of each tribe are rigid, absurd, and often frightfully cruel,[1] and though all outsiders are liable to be regarded as enemies, yet *within those limits* the members live peacefully together—their pursuits, their work, are undertaken in common, thieving and violence are rare, social feeling and community of interest are strong. "In their own bands Indians are perfectly honest. In all my intercourse with them I have heard of not over half-a-dozen cases of such theft. But this wonderfully exceptional honesty extends no further than to the members of his immediate band. To all outside of it, the Indian is not only one of the most arrant thieves in the world, but this quality or faculty is held in the highest estimation." (Dodge, p. 64.) If a man set out on a journey (this among the Kaffirs) "he need not trouble himself about provisions, for he is sure to fall in with some hut, or perhaps a village, and is equally sure of obtaining both food and shelter."[2] "I have lived," says A. R. Wallace in his *Malay Archipelago* (vol. II. p. 460), "with communities in South America and the East, who have no laws or law courts, but the public opinion of the village . . . yet each man scrupulously respects the rights of his fellows, and any infraction of those rights rarely takes place. In such a community all are nearly equal. There are none of those wide distinctions of education and ignorance, wealth and poverty, master and servant, which are the product of our civilisation." Indeed this *community* of life in the early societies, this absence of division into classes, and of the contrast between rich and poor, is now admitted on all sides as a marked feature of difference between the conditions of the primitive and of civilised man.[3]

[1] See Col. Dodge's *Our Wild Indians*.
[2] Wood's *Natural History of Man*. [3] See Appendix.

Lastly, with regard to the mental condition of the Barbarian, probably no one will be found to dispute the contention that he is more easy minded and that his consciousness of Sin is less developed than in his civilised brother. Our unrest is the penalty we pay for our wider life. The missionary retires routed from the savage in whom he can awake no sense of his supreme wickedness. An American lady had a servant, a negro-woman, who on one occasion asked leave of absence for the next morning, saying she wished to attend the Holy Communion? "I have no objection," said the mistress, "to grant you leave; but do you think you *ought* to attend Communion? You know you have never said you were sorry about that goose you stole last week." "Lor' missus," replied the woman, "do ye think I'd let an old goose stand betwixt me and my Blessed Lord and Master?" But joking apart, and however necessary for man's ultimate evolution may be the temporary development of this consciousness of Sin, we cannot help seeing that the condition of the mind in which it is absent is the most distinctively *healthy;* nor can it be concealed that some of the greatest works of Art have been produced by people like the earlier Greeks, in whom it was absent; and could not possibly have been produced where it was strongly developed.

Though as already said, the latest stage of Barbarism, *i.e.*, that just preceding Civilisation, is unrepresented on the earth to-day, yet we have in the Homeric and other dawn-literature of the various nations indirect records of this stage; and these records assure us of a condition of man very similar to, though somewhat more developed than, the condition of the existing races I have mentioned above. Besides this, we have in the numerous traditions of the Golden Age,[1] legends of the Fall, &c., a curious fact which suggests to us that a great number of races in advancing towards Civilisation were con-

[1] See Appendix.

scious at some point or other of having lost a primitive condition of ease and contentment, and that they embodied this consciousness, with poetical adornment and licence, in imaginative legends of the earlier Paradise. Some people indeed, seeing the universality of these stories, and the remarkable fragments of wisdom embedded in them and other extremely ancient myths and writings, have supposed that there really was a general prehistoric Eden-garden or Atlantis; but the necessities of the case hardly seem to compel this supposition. That each human soul however bears within itself some kind of reminiscence of a more harmonious and perfect state of being, which it has at some time experienced, seems to me a conclusion difficult to avoid; and this by itself might give rise to manifold traditions and myths.

II.

HOWEVER all this may be, the question immediately before us—having established the more healthy, though more limited, condition of the pre-civilisation peoples—is, why this lapse or fall? What is the meaning of this manifold and intensified manifestation of Disease—physical, social, intellectual, and moral? what is its place and part in the great whole of human evolution?

And this involves us in a digression, which must occupy a few pages, on the nature of Health.

When we come to analyse the conception of Disease, physical or mental, in society or in the individual, it evidently means, as already hinted once or twice, *loss of unity*. Health, therefore, should mean unity, and it is curious that the history of the word entirely corroborates this idea. As is well known, the words health, whole, holy, are from the same stock; and they indicate to us the fact that far back in the past those who created this group of words had a conception of the meaning of Health very different from ours, and which they embodied unconsciously in the word itself and its strange relatives.

These are, for instance, and among others: heal, hallow, hale, holy, whole, wholesome; German heilig, Heiland (the Saviour); Latin salus (as in salutation, salvation); Greek kalos; also compare hail! a salutation, and, less certainly connected, the root *hal*, to breathe, as in inhale, exhale—French haleine—Italian and French alma and âme (the soul); compare the Latin spiritus, spirit or breath, and Sanskrit âtman, breath or soul.

Wholeness, holiness . . . "if thine eye be single, thy whole body shall be full of light." . . . "thy faith hath made thee *whole*."

The idea seems to be a positive one—a condition of the body in which it is an entirety, a unity—a central force maintaining that condition; and disease being the break-up—or break-down—of that entirety into multiplicity.

The peculiarity about our modern conception of Health is that it seems to be a purely negative one. So impressed are we by the myriad presence of Disease—so numerous its dangers, so sudden and unforetellable its attacks—that we have come to look upon health as the mere absence of the same. As a solitary spy picks his way through a hostile camp at night, sees the enemy sitting round his fires, and trembles at the crackling of a twig beneath his feet—so the traveler thro' this world, comforter in one hand and physic-bottle in the other, must pick his way, fearful lest at any time he disturb the sleeping legions of death—thrice blessed if by any means, steering now to the right and now to the left, and thinking only of his personal safety, he pass by without discovery to the other side.

Health with us is a negative thing. It is a neutralisation of ing dangers. It is to be neither rheumatic nor gouty, consumptive nor bilious, to be untroubled by head-ache, back-ache, heart-ache or any of the "thousand natural shocks that flesh is heir to." These are the realities. Health is then a mere negation of them.

The modern notion, and which has evidently in a very subtle way penetrated the whole thought of to-day, is that the essential fact of life is the existence of innumerable external forces, which, by a very delicate balance and difficult to maintain, concur to produce Man—who in consequence may at any moment be destroyed again by the non-concurrence of those forces. The older notion apparently is that the essential fact of life *is* Man himself; and that the external forces, so-called, are in some way subsidiary to this fact—that they may aid his expression or manifestation, or that they may hinder it, but that they can neither create nor annihilate the Man. Probably both ways of looking at the subject are important; there is a man that can be destroyed, and there is a man that cannot be destroyed. The old words soul and body indicate this contrast; but like all words they are subject to the defect that they are an attempt to draw a line where no line can ultimately be drawn; they mark a contrast where, in fact, there is only continuity—for between the little mortal man who dwells here and now, and the divine and universal Man who also forms a part of our consciousness, is there not a perfect gradation of being, and where (if anywhere) is there a gulf fixed? Together they form a unit, and each is necessary to the other: the first cannot do without the second, and the second cannot get along at all without the first. To use the words of Angelus Silesius (quoted by Schopenhauer), "Ich weiss das ohne mich Gott nicht ein Nu kann leben."

According then to the elder conception, and perhaps according to an elder experience, man to be really healthy must be a unit, an entirety—his more external and momentary self standing in some kind of filial relation to his more universal and incorruptible part—so that not only the remotest and outermost regions of the body, and all the assimilative secretive and other processes belonging thereto, but even the thoughts and passions of the mind itself, stand in direct and clear relationship to it, the final and absolute transparency of the mortal creature. And thus this divinity in each creature, being that which constitutes it and causes it to cohere together, was conceived of as that creature's saviour, healer—healer of wounds of body and wounds of heart—the Man within the man, whom it was not only possible to know, but whom to know and be united with was the alone salvation. This, I take it, was the law of health—and of holiness—as accepted at some elder time of human history, and by us seen as thro' a glass darkly.

And the condition of disease, and of sin, under the same view, was the reverse of this. Enfeeblement, obscuration, duplicity—the central radiation blocked; lesser and insubordinate centres establishing and asserting themselves as against it; division, discord, possession by devils.

Thus in the body, the establishment of an insubordinate centre—a boil, a tumor, the introduction and spread of a germ with innumerable progeny throughout the system, the enlargement out of all reason of an existing organ—means disease. In the mind, disease begins when any passion asserts itself as an independent centre of thought and action. The condition of health in the mind is loyalty to the divine Man within it.[1] But if loyalty to money become an independent centre of life, or greed of knowledge, or of fame, or of drink; jealousy, lust,

[1] No words or theory even of morality can express or formulate this—no enthronement of *any* virtue can take its place; for all virtue enthroned before our humanity becomes vice, and worse than vice.

the love of approbation; or mere following after any so-called virtue for itself—purity, humility, consistency, or what not—these may grow to seriously endanger the other. They are, or should be, subordinates; and though over a long period their insubordination may be a necessary condition of human progress, yet during all such time they are at war with each other and with the central Will; the man is torn and tormented, and is not happy.

And when I speak thus separately of the mind and body, it must be remembered, as already said, that there is no strict line between them; but probably every affection or passion of the mind has its correlative in the condition of the body—though this latter may or may not be easily observable. Gluttony *is* a fever of the digestive apparatus. What is a taint in the mind is also a taint in the body. The stomach has started the original idea of becoming itself the centre of the human system. The sexual organs may start a similar idea. Here are distinct threats, menaces made against the central authority—against the Man himself. For the man must rule or disappear; it is impossible to imagine a man presided over by a Stomach—a walking Stomach, using hands, feet, and all other members merely to carry it from place to place, and serve its assimilative mania. We call such an one, a Hog. [And thus in the theory of Evolution we see the place of the hog, and all other animals, as fore-runners or off-shoots of special faculties in Man, and why the true man, and rightly, has authority over all animals, and can alone give them their place in creation.]

So of the Brain, or any other organ; for the Man is no organ, resides in no organ, but is the central life ruling and radiating among all organs, and assigning them their parts to play.

Disease then, in body or mind, is from this point of view the break-up of its unity, its entirety, into multiplicity. It is the abeyance of a central power, and the growth of insubordinate centres—life in each creature being conceived of as a continual exercise of energy or conquest, by which external or antagonistic forces (and organisms) are brought into subjection and compelled into the service of the creature, or are thrown off as harmful to it. Thus by way of illustration, we find that plants or animals, when in good health, have a remarkable power of throwing off the attacks of any parasites which incline to infest them; while those that are weakly are very soon eaten up by the same. A rose-tree, for instance, brought indoors, will soon fall a prey to the aphis—though when hardened out of doors the pest makes next to no impression on it. In dry seasons when the young turnip plants in the fields are weakly from want of water the entire crop is sometimes destroyed by the turnip fly, which then multiplies enormously; but if a shower or two of rain come before much damage is done the plant will then grow vigorously, its tissues become more robust and resist the attacks of the fly, which in its turn dies. Late investigations seem to show that one of the functions of the white corpuscles in the blood is to devour disease-germs and bacteria present in the circulation—thus absorbing these organisms into subjection to the central life of the body—and that with this object they congregate in numbers toward any part of the body which is wounded or diseased. Or to take an example from society, it is clear enough that if our social life were really vivid and healthy, such parasitic products as the idle shareholder and the policeman above-mentioned would simply be impossible. The material on which they prey would not exist, and they would either perish or be transmuted into useful forms. It seems obvious in fact that life in any organism can only be maintained by some such processes as these—by which parasitic or infesting organisms are either thrown off or absorbed into subjection. To define the nature of the power which thus works towards and creates the distinctive unity of each organism may be difficult, is probably at present impossible, but that some such power exists we can hardly refuse to admit. Probably it is more a subject of the growth of our consciousness, than an object of external scientific investigation.

In this view, Death is simply the loosening and termination of the action of this power—over certain regions of the organism; a process by which, when these superficial parts become hardened and osseous, as in old age, or irreparably damaged, as in cases of accident, the inward being sloughs them off, and passes into other spheres. In the case of man there may be noble and there may be ignoble death, as there may be noble and ignoble life. The inward self, unable to maintain authority over the forces committed to its charge, declining from its high prerogative, swarmed over by parasites, and fallen partially into the clutch of obscene foes, may at last with shame and torment be driven forth from the temple in which it ought to have been supreme. Or, having fulfilled a holy and wholesome time, having radiated divine life and love through all the channels of body and mind, and as a perfect workman uses his tools, so having with perfect mastery and nonchalance used all the materials committed to it, it may quietly and peacefully lay these down, and unchanged (absolutely unchanged to all but material eyes) pass on to other spheres appointed.

And now a few words on the medical aspect of the subject. If we accept any theory (even remotely similar to that just indicated) to the effect that Health is a positive thing, and not a mere negation of disease, it becomes pretty clear that no mere investigation of the latter will enable us to find out what the former is, or bring us nearer to it. You might as well try to create the ebb and flow of the tides by an organised system of mops.

Turn your back upon the Sun and go forth into the wildernesses of space till you come to those limits where the rays of light, faint with distance, fall dim upon the confines of eternal darkness—and phantoms and shadows in the half-light are the product of the wavering conflict betwixt day and night—investigate these shadows, describe them, classify them, record the changes which takes place in them, erect in vast libraries these records into a monument of human industry and research; so shall you be at the end as near to a knowledge and understanding of the sun itself—which all this time you have left behind you, and on which you have turned your back—as the investigators of disease are to a knowledge and understanding of what health is. The solar rays illumine the outer world and give to it its unity and entirety; so in the inner world of each individual possibly is there another Sun, which illumines and gives unity to the man, and whose warmth and light would permeate his system. Wait upon the shining forth of this inward sun, give free access and welcome to its rays of love, and free passage for them into the common world around you, and it may be you will get to know more about health than all the books of medicine contain, or can tell you.

Or to take the former simile: it is the central force of the Moon which acting on the great ocean makes all its waters one, and causes them to rise and fall in timely consent. But take your moon away; hey! now the tide is flowing too far down this estuary! Station your thousands with mops; but it breaks through in channel and runlet! Block it here, but it overflows in a neighboring bay! Appoint an army of swabs there, but to what end? The infinitest care along the fringe of this great sea can never do, with all imaginable dirt and confusion, what the central power does easily, and with unerring grace and providence.

And so of the great (the vast and wonderful) ocean which ebbs and flows within a man—take away the central guide—and not 20,000 doctors, each with 20,000 books to consult and 20,000 phials of different contents to administer, could meet the myriad cases of disease which would ensue, or bolster up into "wholeness" the being from whom the single radiant unity had departed.

Probably there has never been an age, nor any country (except Yankee-land ?) in which disease has been so generally prevalent as in England to-day; and certainly there has never (with the same exception) been an age or country in which doctors have so swarmed, or in which medical science has been so powerful, in apparatus, in learning, in authority, and in actual organisation and number of adherents. How reconcile this contradiction—if indeed a contradiction it be ?

But the fact is that medical science does not contradict disease—any more than laws abolish crime. Medical science—and doubtless for very good reasons—makes a fetish of disease, and dances around it. It is (as a rule) only seen where disease is; it writes enormous tomes on disease; it induces disease in animals (and even men) for the purpose of studying it; it knows, to a marvelous extent, the symptoms of disease, its nature, its causes, its goings out and its comings in; its eyes are perpetually fixed on disease, till disease (for it) becomes the main fact of the world and the main object of its worship. Even what is so gracefully called Hygiene does not get beyond this negative attitude. And the world still waits for its Healer, who shall tell us—diseased and suffering as we are—*what* health is, where it is to be found, whence it flows; and who having touched this wonderful power within himself shall not rest till he has proclaimed and imparted it to men.

No, medical science does not, in the main, contradict disease. The same cause (infidelity and decay of the central life in men) which creates disease and makes men liable to it, creates students and a science of the subject. The Moon[1] having gone from over the waters, the good people rush forth with their mops; and the untimely inundations, and the mops and the mess and the pother, are all due to the same cause.

As to the lodgment of disease, it is clear that this would take place easily in a disorganised system—just as a seditious adventurer would easily effect a landing, and would find insubordinate materials ready at hand for his use, in a land where the central government was weak. And as to the treatment of a disease so introduced there are obviously two methods: one is to reinforce the central power till it is sufficiently strong of itself to eject the insubordinate elements and restore order; the other is to attack the malady from outside and if possible destroy it—(as by doses and decoctions)—independently of the inner vitality, and leaving that as it was before. The first method would seem the best, most durable and effective; but it is difficult and slow. It consists in the adoption of a healthy life, bodily and mental, and will be spoken of later on. The second may be characterised as the medical method, and is valuable, or rather I should be inclined to say, *will* be valuable, when it has found its place, which is to be subsidiary to the first. It is too often however regarded as superior in importance, and in this way, though easy of application, has come perhaps to be productive of more harm than good. The disease may be broken down for the time being, but the roots of it not being destroyed it soon springs up again in the same or a new form, and the patient is as badly off as ever.

The great positive force of Health, and the power which it

[1] It is curious that this word seems to have the same root as the word Man, the original idea apparently being Order, or Measure.

has to *expel* disease from its neighborhood is a thing realised I believe by few persons. But it *has* been realised on earth, and will be realised again when the more squalid elements of our present-day civilisation have passed away.

III.

THE result then of our digression is to show that Health—in body or mind—means unity, integration as opposed to disintegration. In the animals we find this physical unity existing to a remarkable degree. An almost unerring instinct and selective power rules their actions and organisation. Thus a cat before it has fallen (say before it has become a very wheezy fireside pussy!) is in a sense perfect. The wonderful consent of its limbs as it runs or leaps, the adaptation of its muscles, the exactness and inevitableness of its instincts, physical and affectional; its senses of sight and smell, its cleanliness, nicety as to food, motherly tact, the expression of its whole body when enraged, or when watching for prey—all these things are so to speak absolute and instantaneous—and fill one with admiration. The creature is "whole" or in one piece: there is no mentionable conflict or division within it.[1]

Similarly with the other animals, and even with the early man himself. And so it would appear—returning to our subject—that, if we accept the doctrine of Evolution, there is a progression of animated beings—which, though not perfect, possess in the main the attribute of Health—from the lowest forms up to a healthy and instinctive though certainly limited man. During all this stage the central law is in the ascendant,

[1] And with regard to disease, though it is not maintained that among the animals there is anything like immunity from it—since diseases of a more or less parasitic character are common in all tribes of plants and animals—still they seem to be rarer, and the organic instinct of health greater, than in the civilised man.

and the physical frame of each creature is the fairly clean vehicle of its expression—varying of course in complexity and degree according to the point of unfoldment which has been reached. And when thus in the long process of development the inner Man (which has lain hidden or dormant within the animal) at last appears, and the creature consequently takes on the outer frame and faculties of the human being, which are only as they are because of the inner man which they represent; when it has passed through stage after stage of animal life, throwing out tentative types and likenesses of what is to come, and going through innumerable preliminary exercises in special forms and faculties, till at last it begins to be able to wear the full majesty of manhood itself—*then* it would seem that that long process of development is drawing to a close, and that the goal of creation must be within measurable distance.

But then, at that very moment, and when the goal is, so to speak, in sight, occurs this failure of "wholeness" of which we have spoken, this partial break-up of the unity of human nature—and man, instead of going forward any longer in the same line as before, to all appearance *falls.*

What is the meaning of this loss of unity? What is the cause and purpose of this fall and centuries-long exile from the earlier Paradise?

There can be but one answer. It is self-knowledge—(which involves in a sense the abandonment of self). Man has to become conscious of his destiny—to lay hold of and realise his own freedom and blessedness—to transfer his consciousness from the outer and mortal part of him to the inner and undying.

The cat cannot do this. Though perfect in its degree, its interior unfoldment is yet incomplete. The human soul within it has not yet come forward and declared itself; some sheathing leaves have yet to open before the divine flower-bud can be clearly seen. And when at last (speaking as a fool) the cat becomes a man—when the human soul within the creature has climbed itself forward and found expression, transforming the outer frame in the process into that of humanity—(which is the meaning I suppose of the evolution theory)—then the creature, though perfect and radiant in the form of Man, still lacks one thing. It lacks the knowledge of itself; it lacks its own identity, and the realisation of the manhood to which as a fact it has attained.

In the animals consciousness has never returned upon itself. It radiates easily outwards; and the creature obeys without let or hesitation, and with little if any *self*-consciousness, the law of its being. And when man first appears on the earth, and even up to the threshold of what we call civilisation, there is much to show that he should in this respect still be classed with the animals. Though vastly superior to them in attainments, physical and mental, in power over nature, capacity of progress, and adaptability, he still in these earlier stages was like an animal in the unconscious instinctive nature of his action; and on the other hand, though his moral and intellectual structures were far less complete than those of the modern man—as was a necessary result of the absence of self-knowledge—he actually lived more in harmony with himself and with nature,[1] than does his descendant; his impulses,

[1] As to the unity of these wild races with Nature, that is a matter seemingly beyond dispute; their keenness of sense, sensitive to atmospheric changes, knowledge of properties of plants and habits of animals, etc., have been the subject of frequent remark; but beyond this, their strong *feeling* of union with the universal spirit, probably only dimly self conscious, but expressing itself very markedly and clearly in their customs, is most strange and pregnant of meaning. The dances of the Andaman Islanders on the sands at night, the wild festival of the new moon among the Fans and other African tribes, the processions through the forests the chants and dull thudding of drums the torture-dances of

both physical and social, were clearer and more unhesitating; and his unconsciousness of inner discord and sin a great contrast to our modern condition of everlasting strife and perplexity.

If then to this stage belongs some degree of human perfection and felicity, yet there remains a much vaster height to be scaled. The human soul which has wandered darkling for so many thousand of years, from its tiny spark-like germ in some low form of life to its full splendor and dignity in man, has yet to come to the *knowledge* of its wonderful heritage, has yet to become finally individualised and free, to know itself immortal, to resume and interpret all its past lives, and to enter in triumph into the kingdom which it has won.

It has in fact to face the frightful struggle of self-consciousness, or the disentanglement of the true self from the fleeting and perishable self. The animals and man, unfallen, are healthy and free from care, but unaware of what they are; to attain self-knowledge man must fall; he must become less than his true self; he must endue imperfection; division and strife must enter his nature. To realise the perfect Life, to know what, how wonderful it is—to understand that all blessedness and freedom consists in its possession—he must for the moment suffer divorce from it; the unity, the repose of his nature must be broken up, crime, disease and unrest must enter in, and by contrast he must attain to knowledge.

Curious that at the very dawn of the Greek and with it the European civilisation we have the mystic words "Know Thy-

the young Red Indian bravos in the burning heat of the sun; the Dionysiac festivals among the early Greeks; and indeed the sacrificial nature-rites and carnivals and extraordinary powers of second-sight found among all primitive peoples; all these things indicate clearly a faculty which, though it had hardly become self-conscious enough to be what we call religion, was yet in truth the foundation element of religion, and the germ of some human powers which wait yet to be developed.

self" inscribed on the temple of the Delphic Apollo; and that first among the legends of the Semitic race stands that of Adam and Eve eating of the tree of the Knowledge of good and evil! To the animal there is no such knowledge, to the early man there was no such knowledge, and to the perfected man of the future there will be no such knowledge. It is a temporary perversion, indicating the disunion of the present-day man—the disunion of the outer self from the inner—the horrible dual self-consciousness—which is the means ultimately of a more perfect and conscious union than could ever have been realised without it—the death that is swallowed up in victory. "For the first man is of the earth, earthy; but the second man is the Lord from heaven."

In order then, at this point in his Evolution, to advance any farther, Man must first fall; in order to know, he must lose. In order to realise what Health is, how splendid and glorious a possession, he must go through all the long negative experience of Disease; in order to know the perfect social life, to understand what power and happiness to mankind are involved in their true relation to each other, he must learn the misery and suffering which come from mere individualism and greed; and in order to find his true Manhood, to discover what a wonderful power it is, he must first lose it—he must become a prey and a slave to his own passions and desires—whirled away like Phaethon by the horses which he cannot control.

This moment of divorce, then, this parenthesis in human progress, covers the ground of all History; and the whole of Civilisation, and all crime and disease, are only the materials of its immense purpose—themselves destined to pass away as they arose—but to leave their fruits eternal.

Accordingly we find that it has been the work of Civilisation—founded as we have seen on Property—in every way to disintegrate and corrupt man—literally to corrupt—to *break up* the unity of his nature. It begins with the abandonment of the primitive life and the growth of the sense of shame (as in the myth of Adam and Eve.) From this follows the disownment of the sacredness of sex. Sexual acts cease to be a part of religious worship; love and desire—the inner and the outer love—hitherto undifferentiated, now become two separate things. (This no doubt a necessary stage in order for the development of the *consciousness of love*, but in itself only painful and abnormal.) It culminates and comes to an end, as to-day, in a complete divorce between the spiritual reality and the bodily fulfilment—in a vast system of commercial love, bought and sold, in the brothel and in the palace. It begins with the forsaking of the hardy nature-life, and it ends with a society broken down and prostrate, hardly recognisable as human, amid every form of luxury, poverty and disease. He who had been the free child of Nature denies his sonship; he disowns the very breasts that suckled him. He deliberately turns his back upon the light of the sun, and hides himself away in boxes with breathing holes (which he calls houses), living ever more and more in darkness and asphyxia, and only coming forth perhaps once a day to blink at the bright god, or to run back again at the first breath of the free wind for fear of catching cold! He muffles himself in the cast-off furs of the beasts, every century swathing himself in more and more layers, more and more fearfully and wonderfully fashioned, till he ceases to be recognisable as the Man that was once the crown of the animals, and presents a more ludicrous spectacle than the monkey that sits on his own barrel organ. He ceases to a great extent to use his muscles, his feet become partially degenerate, his teeth wholly, his digestion so enervated that he has to cook his food and make pulps of all his victuals, and his whole system so obviously on the decline that at last in the end of time a Kay Robinson arises and prophesies as aforesaid, that he will before long become wholly toothless, bald and toeless.

And so with this denial of Nature comes every form of disease; first delicatesse, daintiness, luxury; then unbalance, enervation, huge susceptibility to pain. With the shutting of himself away from the all-healing Power, man inevitably weakens his whole manhood; the central bond is loosened, and he falls a prey to his own organs. He who before was unaware of the existence of these latter, now becomes only too conscious of them (and this—is it not the very object of the process?); the stomach, the liver and the spleen start out into painful distinctness before him, the heart loses its equable beat, the lungs their continuity with the universal air, and the brain becomes hot and fevered; each organ in turn asserts itself abnormally and becomes a seat of disorder, every corner and cranny of the body becomes the scene and symbol of disease, and Man gazes aghast at his own kingdom—whose extent he had never suspected before—now all ablaze in wild revolt against him. And then—all going with this period of his development—sweep vast epidemic trains over the face of the earth, plagues and fevers and lunacies and world-wide festering sores, followed by armies, ever growing, of doctors—they too with their retinues of books and bottles, vaccinations and vivisections, and grinning death's-heads in the rear—a mad crew, knowing not what they do, yet all unconsciously, doubtless, fulfilling the great age-long destiny of humanity.

In all this the influence of Property is apparent enough. It is evident that the growth of property through the increase of man's powers of production reacts on the man in three ways; to draw him away namely, (1) from Nature, (2) from his true Self, (3) from his Fellows. In the first place it draws him away from Nature. That is, that as man's power over materials increases he creates for himself a sphere and an environment of his own, in some sense apart and different from the great elemental world of the winds and the waves, the woods and the mountains, in which he has hitherto lived. He creates what we call the artificial life, of houses and cities, and shutting himself up in these shuts Nature out. As a growing boy at a certain point, and partly in order to assert his independence, wrests himself away from the tender care of his mother, and even displays—just for the time being—a spirit of opposition to her, so the growing Man finding out his own powers uses them—for the time—even to do despite to Nature, and to create himself a world in which she shall have no part. In the second place the growth of property draws man away from his true Self. This is clear enough. As his power over materials and his possessions increases, man finds the means of gratifying his senses at will. Instead of being guided any longer by that continent and "whole" instinct which characterises the animals, his chief motive is now to use his powers to gratify this or that sense or desire. These become abnormally magnified, and the man soon places his main good in their satisfaction; and abandons his true Self for his organs, the whole for the parts. Property draws the man outwards, stimulating the external part of his being, and for a time mastering him, overpowers the central Will, and brings about his disintegration and corruption. Lastly Property by thus stimulating the external and selfish nature in Man, draws him away from his Fellows. In the anxiety to possess things for himself, in order to gratify his own bumps, he is necessarily brought into conflict with his neighbor and comes to regard him as an enemy. For the true Self of man consists in his organic relation with the whole body of his fellows; and when the man abandons his true Self he abandons also his true relation to his fellows. The mass-Man must rule in each unit-man, else the unit-man will drop off and die. But when the outer man tries to separate himself from the inner, the unit-man from the mass-Man, then the reign of individuality begins—a false and impossible individuality of course, but the only means of coming to the consciousness of the true individuality. With the advent of a Civilisation then founded on Property the unity of the old tribal society is broken up. The ties of blood relationship which were the foundation of the gentile system and the guarantees of the old fraternity and equality become dissolved in favor of powers and authorities founded on mere possession. The growth of Wealth disintegrates the ancient Society; the temptations of power, of possession, &c., which accompany it, wrench the individual from his moorings; personal greed rules; "each man for himself" becomes the universal motto; the hand of every man is raised against his brother; and at last society itself becomes an organisation by which the rich fatten upon the vitals of the poor, the strong upon the murder of the weak. [It is interesting in this connection to find that Lewis Morgan makes the invention of a written alphabet and the growth of the conception of private property the main characteristics of the civilisation-period as distinguished from the periods of savagery and barbarism which preceded it; for the invention of writing marks perhaps better than anything else could do the period when Man becomes *self-conscious*—when he records his own doings and thoughts, and so commences History proper; and the growth of private property marks the period when he begins to sunder himself from his fellows, when therefore the conception of sin (or separation) first enters in, and with it all the long period of moral perplexity, and the denial of that community of life between himself and his fellows which is really of the essence of man's being.]

And then arises the institution of Government.

Hitherto this had not existed except in a quite rudimentary

form. The early communities troubled themselves little about individual ownership, and what government they had was for the most part essentially democratic—as being merely a choice of leaders among blood-relations and social equals. But when the delusion that man can exist for himself alone—his outer and as it were accidental self apart from the great inner and cosmical self by which he is one with his fellows—when this delusion takes possession of him, it is not long before it finds expression in some system of private property. The old community of life and enjoyment passes away, and each man tries to grab the utmost he can, and to retire into his own lair for its consumption. Private accumulations arise; the natural flow of the bounties of life is dammed back, and artificial barriers of Law have to be constructed in order to preserve the unequal levels. Outrage and Fraud follow in the wake of the desire of possession; force has to be used by the possessors in order to maintain the law-barriers against the non-possessors; classes are formed; and finally the formal Government arises, mainly as the expression of such force; and preserves itself, as best it can, until such time as the inequalities which it upholds become too glaring, and the pent social waters gathering head burst through once more and regain their natural levels.

Thus Morgan in his "Ancient Society" points out over and over again that the civilised state rests upon territorial and property marks and qualifications, and not upon a personal basis as did the ancient *gens*, or the tribe; and that the civilised government correspondingly takes on quite a different character and function from the simple organisation of the gens. He says (p. 124), "Monarchy is incompatible with gentilism." Also with regard to the relation of Property to Civilisation and Government he makes the following pregnant remarks (p. 505): "It is impossible to over-estimate the influence of property in the civilisation of mankind. It was the power that brought the Aryan and Semitic nations out of barbarism into civilisation. The growth of the idea of property in the human mind commenced in feebleness and ended in becoming its master passion. Governments and Laws are instituted with primary reference to its creation, protection and enjoyment. It introduced human slavery as an instrument in its production; and after the experience of several thousand years it caused the abolition of slavery upon the discovery that a freeman was a better property-making machine." And in another passage on the same subject, "The dissolution of society bids fair to become the termination of a career of which property is the end and aim; because such a career contains the elements of self-destruction. Democracy is the next higher plane. It will be a revival in a higher form of the liberty, equality and fraternity of the ancient gentes."

The institution of Government is in fact the evidence in social life that man has lost his inner and central control, and therefore must resort to an outward one. Losing touch with the inward Man—who is his true guide—he declines upon an external law, which must always be false. If each man remained in organic adhesion to the general body of his fellows no serious dis-harmony could occur; but it is when this vital unity of the body politic becomes weak that it has to be preserved by artificial means, and thus it is that with the decay of the primitive and instinctive social life there springs up a form of government which is no longer the democratic expression of the life of the whole people; but a kind of outside authority and compulsion thrust upon them by a ruling class or caste.

Perhaps the sincerest, and often though not always the earliest, form of Government is Monarchy. The sentiment of human unity having been already partly but not quite lost, the people choose—in order to hold society together—a man to rule over them who has this sentiment in a high degree. He represents the true Man and therefore the people. This is often a time of extensive warfare and the formation of nations. And it is interesting in this connection to note that the quite early "Kings" or leaders of each nation just prior to the civilisation period were generally associated with the highest religious functions, as in the case of the Roman *rex*, the Greek *basileus*, the early Egyptian Kings, Moses among the Israelites, the Druid leaders of the Britons, and so on.

Later, and as the central authority gets more and more shadowy in each man, and the external attraction of Property greater, so it does in Society. The temporal and spiritual powers part company. The king—who at first represented the Divine Spirit or soul of society, recedes into the background, and his nobles of high degree (who may be compared to the nobler, more generous, qualities of the mind) begin to take his place. This is the Aristocracy and the Feudal Age—the Timocracy of Plato; and is marked by the appearance of large private tenures of land, and the growth of slavery and serfdom—the slavery thus outwardly appearing in society being the symbol of the inward enslavement of the man.

Then comes the Commercial Age—the Oligarchy or Plutocracy of Plato. Honor quite gives place to material wealth; the rulers rule not by personal or hereditary, but by property qualifications. Parliaments and Constitutions and general Palaver are the order of the day. Wage-slavery, usury, mortgages, and other abominations, indicate the advance of the mortal process. In the individual man gain is the end of existence; industry and scientific cunning are his topmost virtues.

Last of all the break-up is complete. The individual loses all memory and tradition of his heavenly guide and counterpart; his nobler passions fail for want of a leader to whom to dedicate themselves; his industry and his intellect serve but to minister to his little swarming desires. This is the era of anarchy—the democracy of Carlyle; the rule of the rabble, and mob-law; caucuses and cackle, competition and universal greed, breaking out in cancerous tyrannies and plutocracies —a mere chaos and confusion of society. For just as we saw in the human body, when the inner and positive force of Health has departed from it, that it falls a prey to parasites which overspread and devour it; so when the central inspiration departs out of social life does it writhe with the mere maggots of individual greed, and at length fall under the dominion of the most monstrous egotist who has been bred from its corruption.

Thus we have briefly sketched the progress of the symptoms of the "disease," which, as said before, runs much (though not quite) the same course in the various nations which it attacks. And if this last stage were really the end of all, and the true Democracy, there were indeed little left to hope for. No words of Carlyle could blast that black enough. But this is no true Democracy. Here in this "each for himself" is no rule of the Demos in every man, nor anything resembling it. Here is no solidarity such as existed in the ancient tribes and primæval society, but only disintegration and a dust-heap. The true Democracy has yet to come. Here in this present stage is only the final denial of all outward and class government, in preparation for the restoration of the inner and true authority. Here in this stage the task of civilisation comes to an end; the purport and object of all these centuries is fulfilled; the bitter experience that mankind had to pass through is completed; and out of this Death and all the torture and unrest which accompanies it, comes at last the Resurrection. Man has sounded the depths of alienation from his own divine spirit, he has drunk the dregs of the cup of suffering, he has literally descended into Hell; henceforth he turns, both in the individual and in society, and mounts deliberately and consciously back again towards the unity which he has lost.[1]

And the false democracy parts aside for the disclosure of the true Democracy which has been formed beneath it—which is not an external government at all, but an inward rule—the rule of the mass-Man in each unit-man. For no outward government can be anything but a make-shift—a temporary hard chrysalis-sheath to hold the grub together while the new life is forming inside—a device of the civilisation-period. Farther than this it cannot go, since no true life can rely upon an external support, and when the true life of society comes all its forms will be fluid and spontaneous and voluntary.

[1] There is another point worth noting as characteristic of the civilisation-period. This is the abnormal development of the abstract intellect in comparison with the physical senses on the one hand, and the moral sense on the other. Such a result might be expected, seeing that abstraction from reality is naturally the great engine of that false individuality or apartness, which it is the object of Civilisation to produce. As it is, during this period man builds himself an intellectual world apart from the great actual universe around him; the "ghosts of things" are studied in books; the student lives indoors, he cannot face the open air—his theories "may prove very well in lecture-rooms, yet not prove at all under the spacious clouds, and along the landscape and flowing currents;" children are "educated" afar from actual life; huge phantom-temples of philosophy and science are reared upon the most slender foundations; and in these he lives defended from actual fact. For as a drop of water when it comes in contact with red-hot iron wraps itself in a cloud of vapor, and is saved from destruction, so the little mind of man, lest it should touch the burning truth of Nature and God and be consumed, evolves at each point of contact a veil of insubstantial thought which allows it for a time to exist apart, and becomes the nurse of its self-consciousness.

IV

AND now, by way of a glimpse into the future—after this long digression what is the route that man will take?

This is a subject that I hardly dare tackle. "The morning wind ever blows," says Thoreau, "the poem of creation is uninterrupted—but few are the ears that hear it." And how can we, gulfed as we are in this present whirlpool, conceive rightly the glory which awaits us? No limits that our present knowledge puts need alarm us; the impossibilities will yield very easily when the time comes; and the anatomical difficulty as to how and where the wings are to grow will vanish when they are felt sprouting!

It can hardly be doubted that the tendency will be—indeed is already showing itself—towards a return to nature and community of human life. This is the way back to the lost Eden, or rather forward to the new Eden, of which the old was only a figure. Man has to undo the wrappings and the mummydom of centuries, by which he has shut himself from the light of the sun and lain in seeming death, preparing silently his glorious resurrection—for all the world like the funny old chrysalis that he is. He has to emerge from houses and all his other hiding places wherein so long ago ashamed (as at the voice of God in the garden) he concealed himself—and Nature must once more become his home, as it is the home of the animals and the angels.

As it is written in the old magical formula: "Man clothes himself to descend, unclothes himself to ascend." Over his spiritual or wind-like body he puts on a material or earthy body; over his earth-body he puts on the skins of animals and other garments; then he hides this body in a house behind curtains and stone walls—which become to it as secondary skins and prolongations of itself. So that between the man and his true life there grows a dense and impenetrable hedge; and what with the cares and anxieties connected with his earth-body and all its skins, he soon loses the knowledge that he is a Man at all, his true self slumbers in a deep and agelong swoon.

But the instinct of all who desire to deliver the divine *imago* within them is, in something more than the literal sense, towards unclothing. And the process of evolution or exfoliation itself is nothing but a continual unclothing of Nature, by which the perfect human Form which is at the root of it comes nearer and nearer to its manifestion.

Thus, in order to restore the Health which he has lost, man has in the future to tend in this direction. Life indoors and in houses has to become a fraction only, instead of the principal part of existence as it is now. Garments similarly have to be simplified. How far this process may go it is not necessary now to enquire. It is sufficiently obvious that our domestic life and clothing may be at once greatly reduced in complexity, and with the greatest advantage—made subsidiary instead of erected into the fetishes which they are. And everyone may feel assured that each gain in this direction is a gain in true life—whether it be the head that goes uncovered to the air of heaven, or the feet that press bare the magnetic earth, or the elementary raiment that allows thro' its meshes the light itself to reach the vital organs. The life of the open air, familiarity with the winds and waves, clean and pure food, the companionship of the animals—the very wrestling with the great Mother for his food—all these things will tend to restore that relationship which man has so long disowned; and the consequent instreaming of energy into his system will carry him to perfections of health and radiance of being at present unsuspected.

Of course, it will be said that many of these things are difficult to realise in our country, that an indoor life, with all its concomitants, is forced upon us by the climate. But if this is to some small—though very small—extent true, it forms no reason why we should not still take advantage of every opportunity to push in the direction indicated. It must be remembered, too, that our climate is greatly of our own creation. If the atmosphere of many of our great towns and of the lands for miles in their neighborhood is devitalised and deadly—so that in cold weather it grants to the poor mortal no compensating power of resistance, but compels him at peril of his life to swathe himself in great-coats and mufflers—the blame is none but ours. It is we who have covered the lands with a pall of smoke, and are walking to our own funerals under it.

That this climate, however, at its best may not be suited to the highest developments of human life is quite possible. Because Britain has been the scene of some of the greatest episodes of Civilisation, it does not follow that she will keep the lead in the period that is to follow; and the Higher Communities of the future will perhaps take their rise in warmer lands, where life is richer and fuller, more spontaneous and more generous, than it can be here.

Another point in this connection is the food question. For the restoration of the central vigor when lost or degenerate, a diet consisting mainly of fruits and grains is most adapted Animal food often gives for the time being a lot of nervous energy—and may be useful for special purposes; but the energy is of a spasmodic feverish kind; the food has a tendency to inflame the subsidiary centres, and so to diminish the central control. Those who live mainly on animal food are specially liable to disease—and not only physically; for their minds also fall more easily a prey to desires and sorrows. In times therefore of grief or mental trouble of any kind, as well as in times of bodily sickness, immediate recourse should be had to the more elementary diet. The body under this diet endures work with less fatigue, is less susceptible to pain, and to cold; and heals its wounds with extraordinary celerity; all of which facts point in the same direction. It may be noted, too, that foods of the seed kind—by which I mean all manner of fruits, nuts, tubers, grains, eggs, etc. (and I may include milk in its various forms of butter, cheese, curds, and so forth), not only contain by their nature the elements of life in their most condensed forms, but have the additional advantage that they can be appropriated without injury to any living creature—for even the cabbage may inaudibly scream when torn up by the roots and boiled, but the strawberry plant *asks* us to take of its fruit, and paints it red expressly that we may see and devour it! Both of which considerations must convince us that this kind of food is most fitted to develop the kernel of man's life.

Which all means cleanness. The unity of our nature being restored, the instinct of bodily cleanness, *both* within and without, which is such a marked characteristic of the animals, will again characterise mankind—only now instead of a blind instinct it will be a conscious, joyous one; dirt being only disorder and obstruction. And thus the whole human being, mind and body, becoming clean and radiant from its inmost centre to its farthest circumference—"transfigured"—the distinction between the words spiritual and material disappears. In the words of Whitman, "objects gross and the unseen soul are one."

But this return to Nature, and identification in some sort with the great cosmos, does not involve a denial or depreciation of human life and interests. It is not uncommonly supposed that there is some kind of antagonism between Man and Nature, and that to recommend a life closer to the latter means mere asceticism and eremitism; and unfortunately this antagonism does exist to-day, though it certainly will not exist for ever. To-day it is unfortunately perfectly true that Man is the only animal who, instead of adorning and beautifying, makes Nature hideous by his presence. The fox and the squirrel may make their homes in the wood and add to its beauty in so doing; but when Alderman Smith plants his villa there, the gods pack up their trunks and depart; they can bear it no longer. The Bushmen can hide themselves and become indistinguishable on a slope of bare rock; they twine their naked little yellow bodies together, and look like a heap of dead sticks; but when the chimney-pot hat and frock-coat appears, the birds fly screaming from the trees. This was the great glory of the Greeks that they accepted and perfected Nature; as the Parthenon sprang out of the limestone terraces of the Acropolis, carrying the natural lines of the rock by gradations scarce perceptible into the finished and human beauty of frieze and pediment, and as, above, it was open for the blue air of heaven to descend into it for a habitation; so throughout in all their best work and life did they stand in this close relation to the earth and the sky and to all instinctive and elemental things, admitting no gulf between themselves and them, but only perfecting their expressiveness and beauty. And some day we shall again understand this which, in the very sunrise of true Art, the Greeks so well understood. Possibly some day we shall again build our houses or dwelling places so simple and elemental in character that they will fit in the nooks of the hills or along the banks of the streams or by the edges of the woods without disturbing the harmony of the landscape or the songs of the birds. Then the great temples, beautiful on every height, or by the shores of the rivers and the lakes, will be the storehouses of all precious and lovely things. There men, women and children will come to share in the great and wonderful common life, the gardens around will be sacred to the unharmed and welcome animals; there all store and all facilities of books and music and art for every one, there a meeting place for social life and intercourse, there dances and games and feasts. Every village, every little settlement, will have such hall or halls. No need for private accumulations. Gladly will each man, and more gladly still each woman, take his or her treasures, except what are immediately or necessarily in use, to the common centre, where their value will be increased a hundred and a thousand fold by the greater number of those who can enjoy them, and where far more perfectly and with far less toil they can be tended than if scattered abroad in private hands. At one stroke half the labor and all the anxiety of domestic caretaking will be annihilated. The private dwelling places, no longer costly and labyrinthine in proportion to the value and number of the treasures they contain, will need no longer to have doors and windows jealously closed against fellow man or mother nature. The sun and air will have access to them, the indwellers will have unfettered egress. Neither man nor woman will be tied in slavery to the lodge which they inhabit; and in becoming once more a part of nature, the human habitation will at length cease to be what it is now for at least half the human race—a prison.

Men often ask about the new Architecture—what, and of what sort, it is going to be. But to such a question there can be no answer till a new understanding of life has entered into people's minds, and then the answer will be clear enough. For as the Greek Temples and the Gothic Cathedrals were built by people who themselves lived but frugally as we should think, and were ready to dedicate their best work and chief treasure to the gods and the common life; and as to-day when we must needs have for ourselves spacious and luxurious villas, we seem to be unable to design a decent church or public building; so it will not be till we once more find our main interest and life in the life of the community and the gods that a new spirit will inspire our architecture. Then when our Temples and Common Halls are not designed to glorify an individual architect or patron, but are built for the use of free men and women, to front the sky and the sea and the sun, to spring out of the earth, companionable with the trees and the rocks, not alien in spirit from the sunlit globe itself or the depth of the starry night—then I say their form and structure will quickly determine themselves, and men will have no difficulty in making them beautiful. And similarly with the homes or dwelling places of the people. Various as these may be for the various wants of men, whether for a single individual or for a family, or for groups of individuals or families, whether to the last degree simple, or whether more or less ornate and complex, still the new conception, the new needs of life, will necessarily dominate them and give them form by a law unfolding from within.

In such new human life then—its fields, its farms, its workshops, its cities—always the work of man perfecting and beautifying the lands, aiding the efforts of the sun and soil, giving voice to the desire of the mute earth—in such new communal life near to nature, so far from any asceticism or inhospitality, we are fain to see far more humanity and sociability than ever before: an infinite helpfulness and sympathy, as between the children of a common mother. Mutual help and combination will then have become spontaneous and instinctive: each man contributing to the service of his neighbor as inevitably and naturally as the right hand goes to help the left in the human body—and for precisely the same reason. Every man—think of it!—will do the work which he *likes*, which he desires to do, which is obviously before him to do, and which he knows will be useful—without thought of wages or reward; and the reward will come to him as inevitably and naturally as in the human body the blood flows to the member which is exerting itself. All the endless burden of the adjustments of labor and wages, of the war of duty and distaste, of want and weariness, will be thrown aside—all the huge waste of work done against the grain will be avoided; out of the endless variety of human nature will spring a perfectly natural and infinite variety of occupations, all mutually contributive; Society at last will be free and the human being after long ages will have attained to deliverance.

This is the Communism which Civilisation has always *hated*, as it hated Christ. Yet it is inevitable; for the cosmical man, the instinctive elemental man accepting and crowning nature, necessarily fulfils the universal law of nature. As to External Government and Law, they will disappear; for they are only the travesties and transitory substitutes of Inward Government and Order. Society in its final state is neither a Monarchy, nor an Aristocracy, nor a Democracy, nor an Anarchy, and yet in another sense it is all of these. It is an Anarchy because there is no outward rule, but only an inward and invisible spirit of life; it is a Democracy because it is the rule of the Mass-man, or Demos, in each unit man; it is an Aristocracy because there are degrees and ranks of such inward power in all men; and it is a Monarchy because all these ranks and powers merge in a perfect unity and central control at last. And so it appears that the outer forms of government which belong to the Civilisation-period are only the expression in separate external symbols of the facts of the true inner life of society.

And just as thus the various external forms of government during the Civilisation-period find their justification and interpretation in the ensuing period, so will it be with the mechanical and other products of the present time; they will be taken up, and find their proper place and use in the time to come. They will not be refused; but they will have to be brought into subjection. Our locomotives, machinery, telegraphic and postal systems; our houses, furniture, clothes, books; our fearful and wonderful cookery, strong drinks, teas, tobaccos; our medical and surgical appliances; high-faluting sciences and philosophies, and all other engines hitherto of human bewilderment, have simply to be reduced to abject subjection to the real man. All these appliances, and a thousand others such as we hardly dream of, will come in to perfect his power and increase his freedom; but they will not be the objects of a mere fetish-worship as now. Man will use them, instead of their using him. His real life will lie in a region far beyond them. But in thus for a moment denying and "mastering" the products of Civilisation, will he for the first time discover their true value, and reap from them an enjoyment unknown before.

The same with the moral powers. As said before, the knowledge of good and evil at a certain point passes away, or becomes absorbed into a higher knowledge. The perception of Sin goes with a certain weakness in the man. As long as there is conflict and division within him, so long does he seem to perceive conflicting and opposing principles in the world without. As long as the objects of the outer world excite emotions in him which pass beyond his control, so long do those objects stand as the signals of evil—of disorder and sin. Not that the objects are bad in themselves, or even the emotions which they excite, but that all through this period these things serve to the man as indications of *his* weakness. But when the central power is restored in man and all things are reduced to his service, it is impossible for him to see badness in anything. The bodily is no longer antagonistic to the spiritual love, but is absorbed into it. All his passions take their places perfectly naturally, and become when the occasions arise the vehicles of his expression. Vices under existing conditions are vices simply because of the inordinate and disturbing influence they exercise, but will cease again to be vices when the man regains his proper command. Thus Socrates having a clean soul in a clean body could drink his boon companions under the table and then go out himself to take the morning air—what was a blemish and defect in them being simply an added power of enjoyment to himself!

The point of difference throughout (being the transference of the centre of gravity of life and consciousness from the partial to the universal man), is symbolized by the gradual resumption of more universal conditions. That is to say that during the civilisation-period, the body being systematically wrapped in clothes, the *head* alone represents man—the little finnikin, intellectual, self-conscious man in contra-distinction to the cosmical man represented by the entirety of the bodily organs. The body has to be delivered from its swathings in order that the cosmical consciousness may once more reside in the human breast. We have to become "all face" again—as the savage said of himself.[1]

Where the cosmic self is, there is no more self-consciousness. The body and what is ordinarily called the self are felt to be only parts of the true self, and the ordinary distinctions of inner and outer, egotism and altruism, etc., lose a good deal of their value. Thought no longer returns upon the local self as the chief object of regard, but consciousness is continually radiant from it, filling the body and overflowing upon external Nature. Thus the Sun in the physical world is the allegory of

[1] See Alonso di Ovalle's *Account of the Kingdom of Chile* in Churchill's *Collection of Voyages and Travels*, 1724.

the true self. The worshipper must adore the Sun, he must saturate himself with sunlight, and take the physical Sun into himself. Those who live by fire and candle-light are filled with phantoms; their thoughts are Will-o'-th'-wisp-like images of themselves, and they are tormented by a horrible self-consciousness.

And when the Civilisation-period has passed away, the old Nature-religion—perhaps greatly grown—will come back. This immense stream of religious life which beginning far beyond the horizon of earliest history has been deflected into various metaphysical and other channels—of Judaism, Christianity, Buddhism, and the like—during the historical period, will once more gather itself together to float on its bosom all the arks and sacred vessels of human progress. Man will once more *feel* his unity with his fellows, he will feel his unity with the animals, with the mountains and the streams, with the earth itself and the slow lapse of the constellations, not as an abstract dogma of Science or Theology, but as a living and ever-present fact. Ages back this has been understood better than now. Our Christian ceremonial is saturated with sexual and astronomical symbols; and long before Christianity existed, the sexual and astronomical were the main forms of religion. That is to say, men instinctively felt and worshipped the great life coming to them through Sex, the great life coming to them from the deeps of Heaven. They deified both. They placed their gods—their own human forms—in sex, they placed them in the sky. And not only so, but wherever they felt this kindred human life—in the animals, in the ibis, the bull, the lamb, the snake, the crocodile; in the trees and flowers, the oak, the ash, the laurel, the hyacinth; in the streams and water-falls, on the mountainsides or in the depths of the sea—they placed them. The whole universe was full of a life which, tho' not always friendly, was *human* and kindred to their own, *felt* by them, not reasoned about, but simply perceived. To the early man the notion of his having a separate individuality could only with difficulty occur; hence he troubled himself not with the suicidal questionings concerning the whence and whither which now vex the modern mind.[1] For what causes these questions to be asked is simply the wretched feeling of isolation, actual or prospective, which man necessarily has when he contemplates himself as a separate atom in this immense universe—the gulf which lies below seemingly ready to swallow him, and the anxiety to find some mode of escape. But when he feels once more that he, that *he* himself, is absolutely indivisibly and indestructibly a part of this great whole—why then there is no gulf into which he can possibly fall; when he is sensible of the fact, why then the *how* of its realisation, tho' losing none of its interest, becomes a matter for whose solution he can wait and work in faith and contentment of mind. The Sun or Sol, visible image of his very Soul, closest and most vital to him of all mortal things, occupying the illimitable heaven, feeding all with its life; the Moon, emblem and nurse of his own reflective thought, the conscious Man, measurer of Time, mirror of the Sun; the planetary passions wandering to and fro, yet within bounds; the starry destinies; the changes of the earth, and the seasons; the upward growth and unfoldment of all organic life; the emergence of the perfect Man, towards whose birth all creation groans and travails—all these things will return to become realities, and to be the frame or setting of his supra-mundane life. The meaning of the old religions will come back to him. On the high tops once more gathering he will celebrate with naked dances the glory of the human form and the great

[1] See Appendix.

processions of the stars, or greet the bright horn of the young moon which now after a hundred centuries comes back laden with such wondrous associations—all the yearnings and the dreams and the wonderment of the generations of mankind—the worship of Astarte and of Diana, of Isis or the Virgin Mary; once more in sacred groves will he reunite the passion and the delight of human love with his deepest feelings of the sanctity and beauty of Nature; or in the open, standing uncovered to the Sun, will adore the emblem of the everlasting splendor which shines within. The same sense of vital perfection and exaltation which can be traced in the early and pre-civilisation peoples—only a thousand times intensified, defined, illustrated and purified—will return to irradiate the redeemed and delivered Man.

In suggesting thus the part which Civilisation has played in history, I am aware that the word itself is difficult to define—is at best only one of those phantom-generalisations which the mind is forced to employ; also that the account I have given of it is sadly imperfect, leaning perhaps too much to the merely negative and destructive aspect of this thousand-year long lapse of human evolution. I would also remind the reader that though it is perfectly true that under the dissolving influence of civilisation empire after empire has gone under and disappeared, and the current of human progress time after time has only been restored again by a fresh influx of savagery, yet its corruptive tendency has never had a quite unlimited fling; but that all down the ages of its dominance over the earth we can trace the tradition of a healing and redeeming power at work in the human breast and an anticipation of the second advent of the son of man. Certain institutions, too, such as Art and the Family (though it seems not unlikely that both of these will greatly change when the special conditions of their present existence have disappeared), have served to keep the sacred flame alive; the latter preserving in island-miniatures, as it were, the ancient communal humanity when the seas of individualism and greed covered the general face of the earth; the former keeping up, so to speak, a navel-cord of contact with Nature, and a means of utterance of primal emotions else unsatisfiable in the world around.

And if it seem extravagant to suppose that Society will ever emerge from the chaotic condition of strife and perplexity in which we find it all down the lapse of historical time, or to hope that the civilisation-process which has terminated fatally so invariably in the past will ever eventuate in the establishment of a higher and more perfect health-condition, we may for our consolation remember that to-day there are features in the problem which have never been present before. In the first place, to-day Civilisation is no longer isolated, as in the ancient world, in surrounding floods of savagery and barbarism, but it practically covers the globe, and the outlying savagery is so feeble as not possibly to be a menace to it. This may at first appear a drawback, for (it will be said) if Civilisation be not renovated by the influx of external Savagery its own inherent flaws will destroy society all the sooner. And there would be some truth in this if it were not for the following consideration: Namely, that while for the first time in History Civilisation is now practically continuous over the globe, now also for the first time can we descry forming in continuous line *within its very structure* the forces which are destined to destroy it and to bring about the new order. While hitherto isolated communisms, as suggested, have existed here and there and from time to time, now for the first time in History both the masses and the thinkers of all the advanced nations of the world are consciously feeling their way towards the establishment of a socialistic and communal life on a vast scale. The present competitive society is more and more rapidly becoming a mere dead formula and husk within which the outlines of the new and *human* society are already discernible. Simultaneously and as if to match this growth, a move towards Nature and Savagery is for the first time taking place from within, instead of being forced upon society from without. The nature-movement begun years ago in literature and art is now among the more advanced sections of the civilised world rapidly realising itself in actual life, going so far even as a denial, among some, of machinery and the complex products of Civilisation, and developing among others into a gospel of salvation by sandals and sunbaths! It is in these two movements—towards a complex human Communism and towards individual freedom and Savagery—in some sort balancing and correcting each other, and both visibly growing up within—tho' utterly foreign to—our present-day Civilisation, that we have fair grounds I think for looking forward to its cure.

APPENDIX.

[SEE p. 9] The following remarks by Mr. H. B. Cotterill on the natives around Lake Nyassa, among whom he lived at a time, 1876-8, when the region was almost unvisited, may be of interest. "In regard of merely 'animal' development and well-being, that is in the delicate perfection of bodily faculties (perceptive), the African savage is as a rule incomparably superior to us. One feels like a child, utterly dependent on them, when traveling or hunting with them. It is true that many may be found (especially amongst the weaker tribes that have been slave-hunted or driven into barren corners) who are half-starved and wizened, but as a rule they are splendid animals. In *character* there is a great want of that strength which in the educated civilised man is secured by the roots striking out into the Past and Future—and in spite of their immense perceptive superiority they feel and acknowledge the superior force of character in the white man.

They are the very converse of the Stoic self-sufficient sage—like children in their 'admiration' and worship of the Unknown. Hence their absolute want of *Conceit*, though they possess self-command and dignity. They are, to those they love and respect, faithful and devoted—their faithfulness and truthfulness are dictated by no 'categorical imperative,' but by personal affection. Towards an enemy they can be, without any conscientious scruples, treacherous and inhumanly cruel. I should say that there is scarcely any possible idea that is so foreign to the savage African mind as that of general philanthropy or enemy-love."

"In *endurance* the African savage beats us hollow (except trained athletes). On one occasion my men rowed my boat with 10 foot oars against the wind in a choppy sea for *25 hours at one go*, across Kuwirwe Bay, about 60 miles. They never once stopped or left their seats—just handed round a handful of rice now and then. I was at the helm all the time—and had enough of it! . . . They carry 80 lbs on their heads for 10 hours through swamps and jungles. Four of my men carried a sick man weighing 14 stones in a hammock for 200 miles, right across the dreaded Malikata Swamp. But for *sudden* emergencies, squalls, etc, they are nowhere."

[See p. 10] "So lovely a scene made easily credible the suggestion, otherwise highly probable, that the Golden Age was no mere fancy of the poets, but a reminiscence of the facts of social life in its primitive organisation of village and house-communities." (J. S. Stuart-Glennie's *Europe and Asia*, ch. I., Servia.)

[See p. 46] "It was only on the up-break of the primitive socialisms that the passionate desire of, and therefore belief in, individual Immortality arose. With an intense feeling, not of an independent individual life, but of a dependent common life, there is no passionate desire of, though there may be more or less of belief in, a continuance after death of individual existence." (*Ibid*, p. 161.)

MODERN SCIENCE:

A CRITICISM.

παντὶ λόγῳ λόγος ἴσος ἀντικεῖται.

IT is one of the difficulties which meet anyone who suggests that modern science is not wholly satisfactory, that it is immediately assumed that the writer is covertly defending what Ingersoll calls the "rib-story," or that he wishes to restore belief in the literal inspiration of the Bible. But, religious controversy apart, and while admitting that Science has done a great work in cleaning away the kitchen-middens of superstition and opening the path to clearer and saner views of the world, it is possible—and there is already a growing feeling that way—that her positive contributions to our comprehension of the order of the universe have in late times been disappointing, and that even her methods are only of limited applicability. After a glorious burst of perhaps fifty years, amid great acclamations and good hopes that the crafty old universe was going to be caught in her careful net, Science, it must be confessed, now finds herself in almost every direction in the most hopeless quandaries; and, whether the rib-story be true or not, has at any rate provided no very satisfactory substitute for it. And the reason of this failure is very obvious. It goes with a certain defect in the human mind, which, as we have pointed out (note, p. 34), necessarily belongs to the Civilisation-period—the tendency, namely, to separate the logical and intellectual part of man from the emotional and instinctive, and to give it a *locus standi* of its own. Science has failed because she has attempted to carry out the investigation of nature from the intellectual side alone—neglecting the other constituents necessarily involved in the problem. She has failed because she has attempted an impossible task; for the discovery of a permanently valid and purely intellectual representation of the universe is simply impossible. Such a thing does not exist.

The various theories and views of nature which we hold are merely the fugitive envelopes of the successive stages of human growth—each set of theories and views belonging organically to the moral and emotional stage which has been reached, and being in some sort the expression of it; so that the attempt at any given time to set up an explanation of phenomena which shall be valid in itself and without reference to the mental condition of those who set it up, necessarily ends in failure; and the present state of confusion and contradiction in which modern Science finds itself is merely the result of such attempt.

Of course this limitation of the validity of Science has been recognised by most of those who have thought about the matter;[1] but it is so commonly overlooked, and latterly the notion has so far gained ground that the "laws" of science are immutable facts and eternal statements of verity, that it may be worth while to treat the subject a little more in detail.

The method of Science is the method of all mundane knowledge; it is that of limitation or actual ignorance. Placed in face of the great uncontained unity of Nature we can only deal with it in thought by selecting certain details and isolating those (either wilfully or unconsciously) from the rest. That is right enough. But in doing so—in isolating such and such details—we practically beg the question we are

[1] See note, p. 81

in search of; and, moreover, in supposing such isolation we suppose what is false, and therefore vitiate our conclusion. From these two radical defects of all intellectual inquiry we cannot escape. The views of Science are like the views of a mountain; each is only possible as long as you limit yourself to a certain stand-point. Move your position, and the view is changed.

Perhaps the word "species" will illustrate our meaning as well as any word; and, in a sense, the word is typical of the method of Science. I see a dog for the first time. It is a fox-hound. Then I see a second fox-hound, and a third and a fourth. Presently I form from these few instances a general conception of "dog." But after a time I see a grey-hound and a terrier and a mastiff, and my old conception is destroyed. A new one has to be formed, and then a new one and a new one. Now I overlook the whole race of civilised dogs and am satisfied with my wisdom; but presently I come upon some wild dogs, and study the habits of the wolf and the fox. Geology turns me up some links, and my conception of dog melts away like a lump of ice into surrounding water. My species exists no more. As long as I knew a few of the facts I could talk very wise about them; or if I limited myself arbitrarily, as we will say, to a study only of animals in England at the present day, I could classify them; but widen the bounds of my knowledge, the area of observation, and all my work has to be done over again. My species is not a valid fact of Nature, but a fiction arising out of my own ignorance or arbitrary isolation of the objects observed.

Or to take an instance from Astronomy. We are accustomed to say that the path of the moon is an ellipse. But this is a very loose statement. On enquiry we find that, owing to perturbations supposed to be produced by the sun, the path deviates considerably from an ellipse. In fact in strict calculations it is taken as being a certain ellipse only for an instant—the next instant it is supposed to be a portion of another ellipse. We might then call the path an irregular curve somewhat resembling an ellipse. This is a new view. But on further enquiry it appears that, while the moon is going round the earth, the earth itself is speeding on through space about the sun—in consequence of which the actual path of the moon does not in the least resemble an ellipse! Finally the sun itself is in motion with regard to the fixed stars, and *they* are in movement too. What then is the path of the moon? No one knows; we have not the faintest idea—the word itself ceases to have any assignable meaning. It is true that if we agree to ignore the perturbations produced by the sun—as in fact we *do* ignore perturbations produced by the planets and other bodies—and if we agree to ignore the motion of the earth, and the flight of the solar system through space, and even the movement of any centre round which that may be speeding, we may then *say* that the moon moves in an ellipse. But this has obviously nothing to do with actual facts. The moon *does* not move in an ellipse—not even "relatively to the earth"—and probably never has done and never will do so. It may be a convenient view or fiction to say that it would do so under such and such circumstances—but it is still only a fiction. To attempt to isolate a small portion of the phenomena from the rest in a universe of which the *unity* is one of Science's most cherished convictions, is obviously self-stultifying and useless.

But you say it can be proved by mathematics that the ellipse would be the path under these conditions; to which I reply that the mathematical proof, though no doubt cogent to the human mind (as at present constituted in most people), is open to the same objection that it does not deal with actual facts. It deals with a mental supposition, *i.e.*, that there are only two bodies acting on each other—a case which never has occurred and never can occur—and then, assuming the law of gravitation (which is just the thing which has to be proved), it arrives at a mental formula, the ellipse. But to argue from this process that the ellipse is really a thing in Nature, and that the heavenly bodies do move or even tend to move in ellipses, is obviously a most unwarrantable leap in the dark. Finally you argue that the leap is warranted because, by assuming that the moon and planets move in ellipses, you can actually foretell things that happen, as for instance the occurrence of eclipses; and in reply to that I can only say that Tycho Brahé foretold eclipses almost as well by assuming that the heavenly bodies moved in epicycles, and that modern astronomers actually do apply the epicycle theory in their mathematical formulæ. The epicycles were an assumption made for a certain purpose, and the ellipses are an assumption made for the same purpose. In some respects the ellipse is a more convenient fiction than the epicycle, but it is no less a fiction.

In other words—with regard to this "path of the moon" (as with regard to any other phenomenon of Nature)—our knowledge of it must be either absolute or relative. But we cannot know the absolute path; and as to the relative, why all we can say is that it does not exist (any more than species exists)—we cannot break up Nature so; it is not a thing in Nature but in our own minds—it is a view and a fiction.

Again, let us take an example from Physics—Boyle's law of the compressibility of gases. This law states that, the temperature remaining constant, the volume of a given quantity of gas is inversely proportional to its pressure. It is a law which has been made a good deal of, and at one time was thought to be true, *i.e.*, it was thought to be a statement of fact. A more extended and careful observation, however, shows that it is only true under so many limitations, that, like the ellipse in Astronomy, it must be regarded as a convenient fiction and nothing more. It appears that air follows the supposed law pretty well, but not by any means exactly except within very narrow limits of pressure ; other gases, such as carbonic acid and hydrogen, deviate from it very considerably —some more than others, and some in one direction and some in the opposite. It was found, among other things, that the nearer a gas was to its liquefying point, the greater was the deviation from the supposed law, and the conclusion was jumped at that the law was true for *perfect* gases only. This idea of a perfect gas of course involved the assumption that gases, as they get farther and farther removed from their liquefying point, reach at last a fixed and stable condition, when no further change in their qualities takes place—at any rate for a very long time—and Boyle's law was supposed to apply to this condition. Since then, however, it has been discovered that there is an ultra-gaseous state of matter, and on all sides it is becoming abundantly clear that the change in the condition of matter from the liquid state to the ultra-gaseous state is perfectly continuous—through all modifications of liquidity and condensation and every degree of perfection and imperfection of gassiness to the utmost rarity of the fourth state. At what point, then, does Boyle's law really apply? Obviously it applies *exactly* at only one point in this long ascending scale—at one metaphysical point—and at every other point it is incorrect. But no gas in Nature remains or can be maintained just at one point in the scale of its innumerable changes. Consequently all we can say is that out of the innumerable different states that gases are capable of, and the innumerable different laws of compressibility which they therefore follow, we could theoretically find one state to which would correspond the law of compressibility called Boyle's law ; and that *if* we could preserve a gas in that state (which we can't) Boyle's law really *would* be true just for that case. In other words, the law is metaphysical. It has no real existence. It is a convenient view or fiction, arising in the first place out of ignorance, and only tenable as long as further observation is limited or wilfully ignored.

This then is the Method of Science. It consists in forming a law or statement by only looking at a small portion of the facts; then when the other facts come in the law or statement gradually fades away again. Conrad Gessner and other early zoologists began by classifying animals according to the number of their horns! Political Economy begins by classifying social action under a law of Supply and Demand. When people believed that the earth was flat they generalised the facts connected with the fall of heavy bodies into a conception of "up and down." These were two opposite directions in space. Heavy bodies took the "downward;" it was their nature. But in time, and as fresh facts came in, it became impossible to group animals any longer by their horns; "up and down" ceased to have a meaning when it was known that the earth was round. Then fresh laws and statements had to be formed. In the last-mentioned case—it being conceived that the earth was the centre of the universe—the new law supposed was that all heavy bodies tended to the centre of the earth as such. This was all right and satisfactory for a while; but presently it appeared that the earth was *not* the centre of the universe, and that some heavy bodies—such as the satellites of Jupiter—did not in fact tend to the centre of the earth at all. Another lump of ignorance (which had enabled the old generalisation to exist) was removed, and a new generalisation, that of universal gravitation, was after a time formed. But it is probable that this law is only conceived of as true thro' our ignorance; nay it is certain that belief in its truth presents the gravest difficulties.

In fact here we come upon an important point. It is sometimes said that, granting the above arguments and the partiality and defectiveness of the laws of Science, still they are approximations to the truth, and as each fresh fact is introduced the consequent modification of the old law brings us *nearer and nearer* to a limit of rigorous exactness which we shall reach at last if we only have patience enough. But is this so? What kind of rigorous statement shall we reach when we have got *all* the facts in? Remembering that Nature is *one*, and that if we try to get a rigorous statement for one set of phenomena (as say the lunar theory) by isolating them from the rest, we are thereby condemning ourselves beforehand to a false conclusion, is it not evident that our limit is at all times infinitely far off? If one knew all the facts relating to a given inquiry except two or three, one might reasonably suppose that one was near a limit of exactness in one's knowledge; but seeing that in our investigation of Nature we only know two or three, so to speak, out of a million, it is obvious that at any moment the fresh law arising from increased experience may completely upset our former calculations. There is a difference between approximating to a wall and approximating to the North Star. In the one case you are tending to a speedy conclusion of your labors, in the other case you are only *going in a certain direction.* The theories of Science generally belong under the second head. They mark the direction which the human mind is taking at the moment in question, but they mark no limits. At each point the *appearance* of a limit is introduced—which becomes, like a mirage in the desert, an object of keen pursuit; but the limit is not really there—it is only an effect of the standpoint, and disappears again after a time as the observer moves. In the case of gravitation there is for the moment an appearance of finality in the law of the inverse square of the distance, but this arises probably from the fact that the law is derived from a limited area of observation only, namely the movements (at great distances from each other) of some of the heavenly bodies.[1] The Cavendish and Schehallien experiments do not show more than that the law at ordinary distances on the earth's surface does not vary *very* much from the above; while the so-called molecular forces compel us (unless we make the very artificial assumption that a variety of attractions and repulsions coexist in matter alongside of, and yet totally distinct from, the attraction of gravitation) to suppose very *great* modifications of the law for small distances. In fact, as we saw of Boyle's law before—the Newtonian law is probably metaphysical—true under certain limited conditions—and the appearance of finality has been given to it by the fact that our observations have been made under such or similar conditions. When we extend our observation into quite other regions of space, the law of the inverse square ceases to appear as even an approximation to the truth—as, for instance, the law of the inverse *fifth* power has been thought to be nearer the mark for small molecular distances.

And indeed the state of the great theories of Science in the present day—the confusion in which the Atomic theory of physics finds itself, the dismal insufficiency of the Darwin theory of the survival of the fittest; the collapse in late times

[1] It is not generally realised how feeble a force gravitation is. It is calculated (Encycl. Brit., Art. Gravitation) that two masses, each weighing 415,000 tons, and placed a mile apart, would exert on each other an attractive force of only one pound. If one, therefore, was as far from the other as the moon is from the earth, their attraction would only amount to $\frac{1}{57,600,000,000\text{th}}$ of a pound. This is a small force to govern the movement of a body weighing 415,000 tons! and it is easy to see that a slight variation in the law of the force might for a long period pass undetected, though in the course of hundreds of centuries it might become of the greatest importance.

of one of the fundamental theories of Astronomy, namely that of the stability of the lunar and planetary orbits; the cataclysms and convulsions which Geology seems just now to be undergoing; the appalling and indeed insurmountable difficulties which attach to the Undulatory theory of Light; the final wreck and abandonment of the Value-theory, the foundation-theory of Political Economy—all these things do not seem to point to very near limits of rigorous exactness! An impregnable theory, or one nearing the limit of impregnability, is in fact as great an absurdity as an impregnable armor-plate. Certainly, given the cannon-balls, you can generally find an armor-plate which will be proof against them; but given the armor-plate, you can always find cannon-balls which will smash it up.

The method of Science, as being a method of artificial limitation or actual ignorance, is curiously illustrated by a consideration of its various branches. I have taken some examples from Astronomy, which is considered the most exact of the physical sciences. Now does it not seem curious that *Astronomy* —the study of the heavenly bodies, which are the most distant from us of all bodies, and most difficult to observe—should yet be the most perfect of the sciences? Yet the reason is obvious. Astronomy is the most perfect science *because we know least about it*—because our ignorance of the actual phenomena is most profound. Situated in fact as we are, on a speck in space, with our observations limited to periods of time which, compared with the stupendous flights of the stars, are merely momentary and evanescent, we are in somewhat the position of a mole surveying a railway track and the flight of locomotives. And as a man seeing a very small arc of a very vast circle easily mistakes it for a straight line, so we are easily satisfied with cheap deductions and solutions in Astronomy, which a more extended experience would cause us to reject.

The man may have a long way to go along his "straight line" before he discovers that it is a curve; he may have much farther to go along his curve before he discovers that it is not a circle; and much farther still to go before he finds out whether it is an ellipse or a spiral or a parabola, or none of these; yet *what* curve it is will make an enormous difference in his ultimate destination. So with the astronomer; and yet Astronomy is allowed to pass as an exact science![1]

Well then, as in Astronomy we get an "exact science" because the facts and phenomena are on such a tremendous scale that we only see a minute portion of them—just a few details so to speak—and our ignorance therefore allows us to

[1] As another instance of the same thing, let me quote a passage from Maxwell's "Theory of Heat," p. 31; the italics are mine: "In our description of the physical properties of bodies as related to heat we have begun with solid bodies, as those which we can *most easily handle*, and have gone on to liquids, which we can keep in open vessels, and have now come to gases, which will escape from open vessels, and which are generally *invisible*. This is the order which is most natural in our first study of these different states. But as soon as we have been made familiar with the most prominent features of these different conditions of matter, the most *scientific* course of study is in the *reverse* order, beginning with gases, on account of the greater simplicity of their laws, then advancing to liquids, the more complex laws of which are much more imperfectly known, and concluding with the little that has been hitherto discovered about the constitution of solid bodies." That is to say that Science finds it easier to work among gases—which are invisible and which we can know little about—than among solids, which we are familiar with and which we can easily handle! This seems a strange conclusion, but it will be found to represent a common procedure of Science—the truth probably being that the laws of gases are not one whit *simpler* than the laws of liquids and solids, but that on account of our knowing so much less about gases it is easier for us to *feign* laws in their case than in the case of solids, and less easy for our errors to be detected.

dogmatise; so at the other end of the scale in Chemistry and Physics we get quasi-exact sciences, because the facts and phenomena are on such a *minute* scale that we overlook *all the details* and see only certain general effects here and there. When a solution of cupric sulphate is treated with ammonia a mass of flocculent green precipitate is formed. No one has the faintest notion of all the various movements and combinations of the molecules of these two fluids which accompany the appearance of the precipitate. They are no doubt very complex. But among all the changes that are taking place, one change has the advantage of being visible to the eye, and the chemist singles that out as the main phenomenon. So chemistry at large consists in a few, very few, facts taken at random as it were (or because they happen to be of such a nature as to be observable) out of the enormous mass of facts really concerned: and because of their fewness the chemist is able to arrange them as he thinks in some order, that is, to generalise about them. But it is certain as can be that he only has to extend the number of his facts, or his powers of observation, to get all his generalisations upset. The same may be said of magnetism, light, heat, and the other physical sciences—but it is not necessary to prove in detail what is sufficiently obvious.

But now, roughly speaking, there is a third region of human observation—a region which does not, like Astronomy (and Geology), lie so far beyond and above us that we only see a very small portion of it; nor, like Chemistry and Physics, so far below us and under such minute conditions of space and time that we can only catch its general effects; but which lies more on a level with man himself—the so-called organic world—the study of man, as an individual and in society, his history, his development, the study of the animals, the plants even, and the laws of life—the sciences of Biology, Sociology, History, Psychology, and the rest. Now this region is obviously that which man knows most of. I don't say that he generalises most about it—but he knows the facts best. For one observation that he makes of the habits and behavior of the stars, or of chemical solutions—for one observation in the remote regions of Astronomy or Chemistry—he makes thousands and millions of the habits and behavior of his fellow-men, and hundreds and thousands of those of the animals and plants. Is it not curious then that in this region he is least sure, least dogmatic, most doubtful whether there be a law or no? Or, rather, is it not quite in accord with our contention, namely that Science, like an uninformed boy, is most definite and dogmatic just where actual knowledge is least.

It will however be replied that the phenomena of living beings are far more complex than the phenomena of Astronomy or Physics—and that is the reason why exact science makes so little way with them. Though man knows many million times more about the habits of his fellow-men than about the habits of the stars, yet the former subject is so many million times more complicated than the latter that all his additional knowledge does not avail him. This is the plea. Yet it does not hold water. It is an entire assumption to say that the phenomena of Astronomy are less complicated than the phenomena of vitality. A moment's thought will show that the phenomena of Astronomy are in reality infinitely complex. Take the movement of the moon: even with our present acquaintance with that subject we know that it has some relation to the position and mass of the earth, including its ocean tides; also to the position and mass of the sun; also to the position and mass of every one of the planets; also of the comets, numerous and unknown as they are; also the meteoric rings; and finally of all the stars! The problem, as everyone knows, is absolutely insoluble even for the shortest period; but when the element of Time enters in, and we consider that to do anything like justice to the problem in an astronomical sense we should have to solve it for at least a million years—during which interval the earth, sun, and other bodies concerned would themselves have been changing their relative positions, it becomes obvious that the whole question is infinitely complex—and yet this is only a small fragment of Astronomy. To debate, therefore, whether the infinite complexity of the movements of the stars is greater or less than the infinite complexity of the phenomena of life, is like debating the precedence of the three persons of the Trinity, or whether the Holy Ghost was begotten or proceeding: we are talking about things which we do not understand.

Nature is one; she is not, we may guess, less profound and wonderful in one department than another; but from the fact that we live under certain conditions and limitations we see most deeply into that portion which is, as it were, on the same level with us. In humanity we look her in the face; there our glance pierces, and we see that she is profound and wonderful beyond all imagination; what we learn there is the most valuable that we can learn. In the regions where Science rejoices to disport itself we see only the skirts of her garments, so to speak, and though we measure them never so precisely, we still see them and nothing more.

There is another point, however, of which much is often made as a plea for the substantial accuracy of the scientific laws and generalisations, namely that they enable us to *predict* events. But this need not detain us long. J. S. Mill in his "Logic" has pointed out—and a little thought makes it obvious—that the success of a prediction does not prove the truth of the theory on which it is founded. It only proves the theory was good enough for that prediction.

There was a time when the sun was a god going forth in his chariot every morning, and there was a time when the earth was the centre of the universe, and the sun a ball of fire revolving round it. In those times men could predict with certainty that the sun would rise next morning, and could even name the hour of its appearance; but we do not therefore think that their theories were true. When Adams and Leverrier foretold the appearance of Neptune in a certain part of the sky, they made a brief prediction to an unknown planet from the observed relations of the movements of the known planets; that does not show however that the grand generalisation of these movements, called the "law of gravitation," is correct. It merely shows that it did well enough for this very brief step—brief indeed compared with the real problems of Astronomy, for which latter it is probably quite inadequate.

Tycho Brahé, excellent astronomer as he was, kept as we saw to the epicycle theory. He imagined that the moon's path round the earth was a fixed combination of cycle and epicycle. Kepler introduced the conception of the ellipse. Later on the motion of the perigee and other deviations compelled the abandonment of the ellipse and the supposition of an endless curve, similar to an ellipse at any one point, and maintaining a fixed mean distance from the earth, but never returning on itself or making a definite closed figure of any kind. Finally the researches of Mr. George Darwin have destroyed the conception of the fixed mean distance, and introduced that of a continually enlarging spiral. Certainly no four theories could well be more distinct from each other than these; yet if an eclipse had to be calculated for next year it would scarcely matter which theory was used. The truth is that the actual problem is so vast that a prediction of a few years in advance only touches the fringe of it so to speak; yet if the fulfilment of the prediction were taken as a proof of the theory in each of these different cases, it would lead in the end to the most hopelessly contradictory results.

The success of a prediction therefore only shows that the theory on which it is founded has had practical value so far as a working hypothesis. As working hypotheses, and as long as they are kept down to brief steps *which can be verified*, the scientific theories are very valuable—indeed we could not do without them; but when they are treated as objective facts—when, for instance, the "law of gravitation"—derived as it is from a brief study of the heavenly bodies—has a universal truth ascribed to it, and is made to apply to phenomena extending over millions of years, and to warrant unverifiable prophecies about the planetary orbits, or statements about the age of the earth and the duration of the solar system—all one can say is that those who argue so are flying off at a tangent from actual facts. For as the tangent represents the direction of a curve over a small arc, so these theories represent the bearing of facts well enough over a small region of observation; but as following the tangent we soon lose the curve, so following these theories for any distance beyond the region of actual observation we speedily part company with facts.[1]

To proceed with a few more words about the general method of Science. Science passes from phenomena to laws, from individual details which can be seen and felt to large generalisations of an intangible and phantom-like character.

[1] All our thoughts, theories, "laws," &c., may perhaps be said to *touch* Nature—as the tangent touches the curve—at a point. They give a direction—and are true—at that point. But make the slightest move, and they all have to be reconstructed. The tangents are infinite in number, but the curve is one. This may not only illustrate the relation of Nature to Science, but also of Art to the materials it uses. The poet radiates thoughts: but he sets no store by them. He knows his thoughts are not true in themselves, but they *touch* the Truth. His lines are the envelope of the curve which is his poem.

That is to say, that for convenience of thought we classify objects. How is this classification effected? It is effected through the perception of identity amid difference. Among a lot of objects I perceive certain attributes in common; this group of common attributes serves, so to speak, as a band to tie these objects together with—into a bundle convenient for thought. I give a name to the band, and that serves to denote any unit of the bundle by. Thus perceiving common attributes among a lot of dogs—as in an example already given—I give the name foxhound to this group of attributes, and thenceforth use the name foxhound to connect these objects by in my mind; again perceiving other common attributes among other similar objects, I invent the word greyhound to denote these latter by. The concept foxhound differs from the objects which it denotes, in this respect that these latter are (as we say) *real* dogs with thousands and thousands of attributes each: one of them has a broken tooth, another is nearly all white, another answers to the name "Sally," and so on; while the concept is only an imaginary form in my mind, with only a few attributes and no individual peculiarities—a kind of tiny G.C.M. arising from the contemplation of a long row of big figures.

Now having created these concepts "foxhound," "greyhound," and a lot of other similar ones, I find that they in their turn have a few attributes in common and thus give rise to a new concept "dog." Of course this "dog" is more of an abstraction than ever, the concept of a concept. In fact the peculiarity of this whole process is that, as sometimes stated, the broader the generalisation becomes the less is its depth; or in other words and obviously, that as the number of objects compared increases, the number of attributes common to them all decreases. Ultimately as we saw at the beginning, when a sufficient number of objects are taken in, the concept ("dog" or whatever it may be) fades away and ceases to have any meaning. This therefore is the dilemma of Science and indeed of all human knowledge, that in carrying out the process which is peculiar to it, it necessarily leaves the dry ground of reality for the watery region of abstractions, which abstractions become ever more tenuous and ungraspable the farther it goes, and ultimately fade into mere ghosts. Nevertheless the process is a quite necessary one, for only by it can the mind deal with things.

To dwell for a moment over this last point: it is clear that every object has relation to every other object—exists in fact only in virtue of such relation to other objects; it has therefore an infinite number of attributes, the mind consequently is powerless to deal with such object—it cannot by any possibility think it. In order to deal with it, the mind is forced to single out a *few* of its attributes (the *method of ignorance* or abstraction already alluded to)—that is a few of its relations to other objects, and to think them first. The others it will think afterwards—all in good time. In thus stripping or abstracting the great mass of its attributes from our object, and leaving only a few, which it combines into a concept, the mind practically abandons the real article and takes up with a shadow; but in return for this it gets something which it can handle, which is light to carry about, and which like paper-money, *for the time and under certain conditions* does really represent value. The only danger is lest it—the mind—carried away by the extensive applicability of the partial concept which it has thus formed, should credit it with an actual value—should project it on the background of the external world and ascribe to it that reality which belongs only to objects themselves, *i.e.*, to things embodying an *infinite* range of attributes.

The peculiar method of Science is now clear to us, and can

be abundantly illustrated from modern results. Our experience consists in sensations, we feel the weight of heavy bodies, we see them fall when let go, we have sensations of heat and cold, light and darkness, and so forth. But these sensations are more or less local and variable from man to man, and we naturally seek to find some common measure of them, by which we can talk about and describe them *exactly*, and independently of the peculiarities of individual observers. Thus we seek to find some common phenomenon which underlies (as we say) the sensations of heat and cold, or of light and darkness, or something which explains (*i.e.* is always present in) the case of falling bodies—and to do this we adopt the method of generalisation above described, *i.e.*, we observe a great number of individual cases and then see what qualities or attributes they have in common. So far good. But it is just here that the fallacy of the ordinary scientific procedure comes in; for, forgetting that these common qualities are mere abstractions from the real phenomena we credit *them* with a real existence, and regard the actual phenomena as secondary results, "effects" or what-not of these "causes." This in plain language is putting the cart before the horse—or rather the shadow before the man. Thus finding that a vast number of variously shaped and colored bodies tend to fall towards the earth, we erect this common attribute of falling into an independent existence which we call "attraction" or "gravitation"—and ultimately posit a universal gravitation *acting* on all bodies in Nature!—or finding that a number of different substances, such as water, air, wood, &c., convey to us the sensation we call sound, and that in all these cases the common element is vibration, we detach the attribute vibration, credit it with a separate existence, and speak of it as the cause of sound. But though we may thus *think* of the shadow as separate from the man, the shadow cannot *be* separate from the man; and tho' we may try to think of the falling or the vibration as separate from the wood or the stone, such falling and vibration cannot exist apart from these and other such materials, and the effort to speak of it as so existing ends in mere nonsense. More strange still is the fatuity, when, as in the case of the undulatory Theory of light or the Atomic theory of physics, the concepts thus erected into actualities are composed of purely imaginary attributes—of which no one has had any experience—an undulatory ether in the one case, a hard and perfectly elastic atom in the other. The total result is of course—just what we see—Science landing itself in pure absurdities in every direction. Beginning by detaching the attribute of falling from the bodies that fall—beginning that is by an abstraction, which of course is also a falsity—it generalises and generalises this abstraction till at last it reaches a perfectly generalised absurdity and thing without any meaning—the law of gravitation. The statement that "every particle in the universe attracts every other particle with a force proportional to the mass of the attracting particle and inversely proportional to the square of the distance between the two" is devoid of meaning—the human mind can give no definite meanings to the words "mass," "attract," and "force," which do not stultify each other. The law in every way baffles intelligence. Newton, who invented it, declared that no philosophic mind would suppose that bodies could thus act on one another "without the mediation of anything else by and through which their action might be conveyed;" scientific men to-day are fain to see that a material mediation of this kind would only make the law still more unintelligible than it is, while, on the other hand, an immaterial mediation or a fourth-dimensional mediation, such as some propose, would simply remove the problem out of the regions of scientific analysis. Again the form of the law is declared to be the inverse square of the distance; but this is the law by the nature of space itself of any perfect radiation, and if true of gravitation involves the conclusion that that radiation of force (whatever its nature may be) takes place without loss or dissipation of any kind. This would make gravitation absolutely unique among phenomena. More than this, its propagation is supposed to be *instantaneous* over the most enormous distances of space, and to take place always unhindered and unretarded whatever be the number or the nature of the bodies between! What can be more clear than that the law is simply metaphysical—a projection into a monstrous universality and abstraction, of partially understood phenomena in a particular region of observation—a Brocken-shadow on the background of Nature of the observer's own momentary attitude of thought?

Again, the undulatory theory of Light. Studying the phenomena of a vast number of colored and bright bodies, Science finds that it can think about these phenomena—can generalise and tie them into bundles best by *assuming* that the bodies are all in a state of vibration; a vibration so minute that (unlike the vibrations connected with Sound) it cannot be directly perceived. So far good. There is no harm in the assumption of vibration as long as it is understood to be a mere assumption for a temporary convenience of thought. But now Science goes farther than this, and not only supposes a common attribute to all visible bodies, but credits this common attribute with a real existence independent of the visible bodies in which it was supposed to inhere—and makes this the *cause* of their visibility! Obviously now a common and universal medium is required for this common and universal assumed vibration (just as Newton required a medium for his universal "falling")—and so, hey presto! we have the Undulatory Ether. And having got it we find that to fulfil our requirements it must have a pressure of 17 million million pounds on the square inch, and yet be so rare and tenuous as not to hinder the lightest breath of air; that while it is thus rare enough to surpass all our powers of direct scrutiny, its vibrations must yet be capable of agitating and breaking up the solidest bodies; that it must pass freely through some dense and close structures like glass, and yet be excluded by some light and porous, like cork, and so on and on! In fact we find that it is unthinkable. Against this adamantine, impalpable Ether, as against this instantaneous, untranslateable gravitation, Science bangs its devoted head in vain. Having created these absurdities by the method of "personification of abstractions"[1] or the "reification of concepts,"[2] it seriously and in all good faith tries to understand them; having dressed up its own Mumbo Jumbo (which it once jeered at religion for doing) it piously shuts its eyes and endeavors to believe in it.

The Atomic Theory[3] affords a good example of the "method of ignorance." When we try to think about material objects generally—to generalise about them—that is, to find some attribute or attributes common to them, we are at first puzzled. They present such an immense variety. But after a time, by dint of stripping off or abstracting all such attributes or qualities as we think we perceive in one body and not in another—as for example, redness, blueness, warmth, saltness, life, intelligence, or what not—we find an attribute left, namely resistance to touch, which is common to *all* material bodies. This quality in the body we call "mass," and since it is only known by motion, mass and motion become correlative attributes which we find useful to class bodies by, not because they represent the various bodies particularly well, but because they are found in all bodies; just as you might class people by their boots—not because boots are a very

[1] J. S. Mill. [2] Stallo.

[3] See Stallo's excellent *Concepts of Modern Physics*.

valuable method of classification, but simply because every one wears boots of one kind or another. So far there is no great harm done. But now having by the method of ignorance *thought away* all the qualities of bodies, except the two correlatives of mass and motion, we set about to *explain* the phenomena of Nature generally by these two "thinks" that are left. We credit these "thinks" (mass and motion) with an independent existence and proceed to derive the rest of phenomena from them. The proceeding of course is absurd, and ends by exposing its own absurdity. Thinking of mass and motion as existing in the various bodies *apart* from color, smell, and so forth—which of course is not the case—we combine the two attributes into one concept, the atom, which we thus assume to exist in all bodies. The atom has neither color, smell, warmth, taste, life or intelligence; it has only mass and motion; for it came by the method of divesting our thought of everything *but* mass and motion. It is a projection of a "think" upon the background of nature. And it is an absurdity. No such thing exists in all the wide universe as mass and motion divested from color, smell, warmth, life and intelligence. The atom is unthinkable. It is perfectly hard and it is perfectly elastic—which is the same as saying that it bends and it doesn't bend at the same time; it has form, and it hasn't form; it has affinities and yet is perfectly indifferent. To justify to men the ways of their Mumbo Jumbo has sorely exercised the votaries of the Atom. One philosopher says that it is mere matter, passive, exercising no force but resistance; another says that it is a centre of force, without matter; a third suggests that it is not itself matter, but only a vortex in other matter! All agree that it is not an object of sense, and there remains no conclusion but that it is nonsense![1]

[1] See, for instance, the last new thing in this style—the Helmholtz

And so on in all directions. Human thought flying off at its tangents from Nature lands itself in infinite nothings afar off, poor ghostly skeletons and abstractions from Nature—which indeed is all right, for human thought as yet can only see ghosts and not realities; but let there be no mistake, let these ghosts not be mistaken for realities—for they are not even compatible with each other. The Atom that suits the physicist does not suit the chemist. The Ether that does for the vehicle of Light will not do for the vehicle of universal Gravitation.

It would be hardly worth while entering into these criticisms, were it not evident that Science in modern times, either tacitly or explicitly, has been seeking, as I said at the beginning, to enounce facts independent of Man, the observer. Seeing that the ordinary statements of daily life are obviously inexact and relative to the observer—charged with human sensation in fact—Science has naturally tried to produce something which should be exact and independent of human sensation; but here it has of course condemned itself beforehand to failure; for no statement of isolated phenomena or groups of phenomena *can* be exact except by the method of ignorance aforesaid, and no statement obviously can be really independent of human sensation. When a man says *It is cold*, his statement, it must be confessed, is deplorably human and vague. *It*—what is that? *Is*—do you mean *is*? or do you

molecule as improved upon by Sir William Thomson; it is described as follows; "A heavy mass connected by massless springs with a massless enclosing shell; or there may be several shells enclosing each other connected by springs with a dense mass in the centre (far more dense than the ether)." It is not, of course, seriously maintained that this nonsensical creation exists—but that if it did exist it would account for certain unexplained phenomena in the dispersion of light, &c.

mean *feels, appears?* *Cold*—in what sense? Cold to yourself, or to other people, or to polar bears, or by the thermometer? And so on. Science therefore steps in with an air of authority and sets him right. It says *the temperature is* 30° *Fahrenheit*, as if to settle the matter. But does this really settle the matter? *Temperature*—who knows what that is? What is the scientific definition of it? I find (Clerk-Maxwell's Theory of Heat, p. 2.) "the temperature of a body is a quantity which indicates how hot or how cold the body is." This sounds very much like saying, "the color of a body is a quantity which indicates how blue, red, or yellow the body is." It does not bring us much farther on our way. But in the next paragraph Maxwell shows the object of his definition (which of course is only preliminary) by saying, "By the use, therefore, of the word temperature, we fix in our minds the conviction that it is possible not only to feel, but to *measure*, how hot a body is." That is to say he clearly maintains that it is possible to find an absolute standard of hotness or coldness—or rather of the unknown thing called temperature—outside of ourselves and independent of human sensation. When the man said he was cold he was probably just describing his own sensations, but here Science indicates that it is in search of something which has an independent existence of its own, and which therefore when found we can measure exactly and once for all. What then is that thing? *What* is temperature? say, what is it?

We cudgel our brains in vain. Perhaps the remainder of the sentence will help us. "The temperature is 30° Fahrenheit." "The unknown thing is thirty degrees." When then is a degree? That is the next question. When the Theory of Heat went out from sensation and left it behind, one of its first landing places was in the expansion of liquids—as in thermometer tubes. Here for some time was thought to be a satisfactory register of "temperature." But before long it became apparent that the degree—Fahrenheit, Réaumur, or what-not—was an entirely arbitrary thing, also that it was not the *same*[1] thing at one end of the scale as the other, and finally that the scale itself had no starting point! This was awkward, so a move was made to the air thermometer, and there was some talk about an absolute zero and absolute temperatures; it was thought that the Unknown thing showed itself most clearly and simply in the expansion of air and other gases, and that the "degree" might fairly be measured in terms of this expansion. But in a little time this kind of thermometer—chiefly because no gas turned out to be "theoretically perfect"—broke down, absolute zero and all, and another step had to be made—namely, to the dynamical theory. It was announced that the Unknown thing might be measured in terms of mechanical energy, and Joule at Manchester proclaimed that the work done by any quantity of water falling there a distance of 772 feet is capable of raising that water one degree Fahrenheit.[2] Here seemed something definite. To measure temperature by mass and velocity, to measure a degree by the flight of a stone, or the heat in the human body by the fall of a factory chimney—if rather roundabout and elusive of the main question—seemed at any rate promising of exact results! Unfortunately the difficulty was to pass from the theory to its application. The complicated nature of the problem, the "imperfection" of the gases and other bodies under consideration, the latent and specific heats to be allowed for, the elusive nature of heat in experiment,

[1] The very fact alone that the degrees on a thermometer are *equal* space divisions shows that they must bear a *varying* relation to the total volume of liquid as that expands from one end of the tube to the other.

[2] A statement obviously applying—from what has been already said—at only one point in the scale.

and the variable value of the degree itself—all render the conclusions on this subject most precarious; and the general equations connecting the Fahrenheit or other temperatures with a thermo-dynamic scale—while they become so unwieldy as to be practically useless—are themselves after all only approximate.

Finally, to give a last form to the mechanical theory of heat, the conception of flying atoms or molecules was introduced, and a number of neat generalisations were deduced from dynamical considerations. Of course it was inevitable, having once started with a mechanical theory, that one should arrive at the Atom some time or other—and (from what has already been said) it was also inevitable that the result should be unsatisfactory. It is sufficient to say that the molecular theory of heat is *not* in accordance with facts. Such things as the law of Charles and the law of Boyle, which according to it should be strictly accurate and of general application, are known to be true only over a most limited range. This failure of the theory may be said to arise partly from its being pursued by the statistical method; but if, on the other hand, we were to try and follow out the individual movement of each molecule we should be landed in a problem far exceeding in complexity the wildest flights of Astronomy—and should have exchanged for the original difficulty about "temperature" a difficulty far greater.

The result of all this has been that notwithstanding the talk about energy and atoms, Science has sadly to confess that it can still give no valid meaning to the word temperature: the unknown thing is still unknown, the independent existence round the corner still escapes us. By the very effort to arrive at something independent of human sensation, Science has, in a roundabout way, arrived at an absurdity. When the man said he was cold, his statement—deplorably vague as it was—had some meaning: he was describing his feelings, or possibly he had seen some snow or some ice on the road; but when, in the endeavor to leave out the human and to say something absolute, Science declared that the temperature was thirty degrees, it committed itself to a remark which possibly was exact in form, but to which it has never given and never can give any meaning.[1]

Similarly with other generalities of Science: the "law" of the Conservation of Energy, the "law" of the Survival of the Fittest—the more you think about them the less possible is it to give any really intelligible sense to them. The very word Fittest really begs the question which is under consideration, and the whole Conservation law is merely an attenuation of the already much attenuated "law" of Gravitation. The Chemical Elements themselves are nothing but the projection on the external world of concepts consisting of three or four attributes each: they are not more real, but very much less real than the individual objects which they are supposed to account for; and their "elementary" character is merely fictional. It probably is in fact as absurd to speak of pure carbon or pure gold, as of a pure monkey or a pure dog. There are no such things, except as they may be arrived at by arbitrary definition and the method of ignorance.

In the search for exactness then Science has been continually led on to discard the human and personal elements in phenomena, in the hope of finding some residuum as it were behind them which should not be personal and human but absolute

[1] I am not, of course, here arguing against the use of thermometers or other instruments for practical purposes. This is certainly *the* legitimate field of Science. But, as in the case of *prediction* before mentioned, the exactness of certain practical results obtained is a very different matter from the truth of the generalities which are supposed to underlie these results. In using a thermometer you need not even mention the word "temperature."

and invariable. And the tendency has been (hitherto) in all the sciences to get rid of such terms as blue, red, light, heavy, hot, cold, concord, discord, health, vitality, right, wrong, &c., and to rely on any less human elements discoverable in each case; as for instance in Sound, to deal less and less with the judgments and sensations of the ear, and to rely more and more on measurements of lengths of strings, numbers of vibrations, &c. Each science has been (as far as possible) reduced to its lowest terms. Ethics has been made a question of utility and inherited experience. Political Economy has been exhausted of all conceptions of justice between man and man, of charity, affection, and the instinct of solidarity; and has been founded on its lowest discoverable factor, namely self-interest. Biology has been denuded of the force of personality in plants, animals, and men; the "self" here has been set aside, and the attempt made to reduce the science to a question of chemical and cellular affinities, protoplasm, and the laws of osmose. Chemical affinities, again, and all the wonderful phenomena of Physics are emptied down into a flight of atoms; and the flight of atoms (and of astronomic orbs as well) is reduced to the laws of dynamics—which the student sitting in his chamber may write down on a piece of paper. Thus the idea, formulated by Comte, of a great scale of sciences arising from the simplest to the most complex, has tacitly underlain modern scientific work. It—Science—has sought to "explain" each stage by reference to a lower stage—"blueness" by vibrations, and vibrations by flying atoms—the human always by the sub-human. Going out from humanity dissatisfied, it has wandered through the animal and vegetable kingdoms, through the regions of Chemistry and Physics, into that of Mechanics. "Here at last, in Mechanics, is something outside humanity, something exact in itself, something substantial," it has said. "Let us build again on this as on a foundation, and in time we shall find out what humanity is." This I say has been the dream of Modern Science; yet the fallacy of it is obvious. We have not got outside the human, but only to the outermost verge of it. Mass and motion, which in this process are taken to be real entities and the first progenitors of all phenomena, are simply the last abstractions of sensible experience, and our emptiest concepts. The *material* explanation of the universe is simply an attempt to account for phenomena by those attributes which appear to us to be common to them all —which is, as said before, like accounting for men by their boots;—it may be possible to get an exact formula this way, but its contents have little or no meaning.

The whole process of Science and the Comtian classification of its branches—regarded thus as an attempt to explain Man by Mechanics—is a huge vicious circle. It professes to start with something simple, exact, and invariable, and from this point to mount step by step till it comes to Man himself; but indeed it starts with Man. It plants itself on sensations low down (mass, motion, &c.), and endeavors by means of them to explain sensations high up, which reminds one of nothing so much as that process vulgarly described as "climbing up a ladder to comb your hair." In truth Science has never left the great world, or cosmos, of Man, nor ever really found a *locus standi* without it; but during the last two or three centuries it has gone in this *direction*, outwards, continually. Leaving the central basis and facts of humanity as too vast and unmanageable, and also as apparently variable from man to man and therefore affording no certain consent to work upon, it has wandered gradually outwards, seeking something of more definite and universal application. Discarding thus one by one the interior phases of sensation—as the sense of personal relationship, the sense of justice, duty, fitness in things or what-not (as too uncertain, or perhaps developed to an unequal degree in different persons, embryonic in one and matured in another), drifting past the more specialised bodily senses, of color, sound, taste, smell, &c., as for similar reasons unavailable—Science at last in the primitive consciousness of muscular contraction and its abstraction "mass" or "matter" comes to a pause. Here in this last sense, common probably to man and the lowest animals, it finds its widest, most universal ground—its farthest limit from the Centre. It has reached the outermost shell, as it were, of the great Man-cosmos. Even this shell is partially human; it is not entirely osseous, and so far not entirely exact and invariable; but Science can go no farther—and there, for the present, it may remain!

Some day perhaps, when all this showy vesture of scientific theory (which has this peculiarity that only the learned can *see* it) has been quasi-completed, and Humanity is expected to walk solemnly forth in its new garment for all the world to admire—as in Anderssen's story of the Emperor's New Clothes—some little child standing on a door-step will cry out: "But he has got nothing on at all," and amid some confusion it will be seen that the child is right.

NOTE.

"I fear I have very imperfectly succeeded in expressing my strong conviction that, before a rigorous logical scrutiny, the Reign of Law will prove to be an unverified hypothesis, the Uniformity of Nature an ambiguous expression, the certainty of our scientific inferences to a great extent a delusion." (Stanley Jevons. *Principles of Science*, p. ix.)

F

THE SCIENCE OF THE FUTURE:

A FORECAST.

Once let that [the human ideal] slip out of the thought, and science is of no more use than the invocations in the Egyptian papiri.

RICHARD JEFFERIES.

IT would appear then, from the preceding paper, that in some sense a mistake has been made in the method of modern scientific work; not that the vast amount of labor expended in it has been altogether wasted, for in return for this there is a mass of practical results and detailed observations to show; but that in attempting to solve the problem of science by the intellect alone, a radical mistake has been made which *could* only land us in absurdity, and that this mistake has for the time being also vitiated the results that have been attained. For—in reference to this last point—the divorce of the intellectual from the emotional has caused a great portion of our scientific observations to become merely pedantic and trifling; while it has turned the practical results—as industrial and military machinery, &c.—into engines of evil as often as into engines of good.

Science in searching for a permanently valid and purely intellectual representation of the universe has, as already said, been searching for a thing which does not exist. The very facts of Nature, as we call them, are at least half feeling. If we try to clean the feeling out of a fact and to produce a statement which shall be devoid of the human or sense element, it simply amounts to cleaning the meaning out; and though our resulting statement may be exact it is nugatory and of no value. We might as well try to take the clay out of a brick. It must never be forgotten that the logical processes—important as they are—cannot stand by themselves, have no standing ground of their own. They presuppose assumptions and are the expression of things that are unreasoning, perhaps illogical. The strictest logic is a mere hooking together of links in a chain, and the last link is of no use—you can put no stress on it—unless the first is secured somewhere. The strength of the intellectual chain is no greater than that of the staple from which it hangs—and that is a human feeling. The strength of Euclid is no greater than that of the axioms—and *they* are feelings; they are unreasoning statements of which all that we can say is, "I *feel* like that." In fact, all the propositions of Geometry are nothing but the analysis and elaborate expression, so to speak, of these primary convictions—and the Geometry-structure stands and falls with them. There is no such thing as intellectual truth—that is, I mean, a truth which can be stated as existing apart from feeling. If, for instance, a proposition in Geometry can be really shown to be based on the axioms, it is true, not intellectually or absolutely, but as an expression of my primary Geometrical sense; and if my giving a few pence to a crossing sweeper is based not on a mere impression of duty, or an anxiety to appear charitable, or wish to escape his importunity, but on genuine regard for the man, then it is true, not in any absolute signification, but just as an expression of what it professes to represent—namely my primary sense of humanity. Indeed the truest truth is that which is the expression of the deepest feeling, and if there is an absolute truth it can only be known and expressed by him who has the absolute feeling or Being within himself.

This being so—and the nature of the intellectual processes being, like the links in a chain, transitional—it becomes obvious that the intellectual results may figure as a *means* but never as an end in themselves. To hang any weight of reliance on them in the latter sense is like the Chinese Trick—described by Marco Polo—of throwing a rope's end up in the air and then climbing up the rope. Hence it appears that our scientific theories are perfectly legitimate as long as they are formed as a means towards *practical* applications. In that sense they are transitional; they are formed not as substantial truths but merely as links in a chain towards some definite practical result. For this purpose we may form whatever theories are convenient: if we are calculating the strength of bridges, we may adopt what generalisations we like concerning mechanical structure, as long as they give us actual and practical results; if we are predicting eclipses, we may make use of any theory that will do. The theory does not matter as long as it hauls the practical result after it, just as it does not matter whether your cable is of iron or hemp or silk as long as you can get your ship into dock with it. In this sense our Modern Science is, I conceive, admirable. For practical results and brief predictions it affords a quantity of useful generalisations—shorthand notes and conventional symbols and pocket summaries of phenomena—which bear about the same relation to the actual world that a map does to the country it is supposed to represent. It cannot be said to have any resemblance to the real thing—but when you understand the principle on which it is formed it is exceedingly useful for finding your way about. As long as Science therefore keeps the practical end in view, and starting from sense seeks to return to sense again, its intermediate theorising is perfectly legitimate; but the moment it credits its theory with a positive and authoritative existence, as an actual representation of facts—and endeavors to pass by means of it into unverifiable and abstract regions, as of invisible germs or atoms, or far distances of space, or the remote past or future—it is simply throwing its rope's end into the sky and trying to climb up!

That "the wish is father to the thought" is in its wide sense profoundly true. In the individual, feeling precedes thinking—as the body precedes the clothes. In history, the Rousseau precedes the Voltaire. There is, I believe, a physiological parallel; for behind the brain and determining its action stands the great sympathetic nerve—the organ of the emotions. In fact here the brain appears as distinctly transitional. It stands between the nerves of sense on the one hand and the great sympathetic on the other.

Change the feeling in an individual, and his whole method of thinking will be revolutionised; change the axiom or primary sensation in a science, and the whole structure will have to be re-created. The current Political Economy is founded on the axiom of individual greed; but let a new axiomatic emotion spring up (as of justice or fair play instead of unlimited grab), and the base of the science will be altered, and will necessitate a new construction.

So when people argue (on politics, morality, art, &c.) it will generally be found that they differ at the *base;* they go out, perhaps quite unconsciously, from different axioms and hence they *cannot* agree. Occasionally of course a strict examination will show that, while agreeing at the base, one of them has made a false step in deduction; in that case his thought does *not* represent his primary feeling, and when this is pointed out he is forced to alter it. But more often it is found that the difference lies deep down at a point beyond the reach of reason; and they disagree to the end. In this case neither is right and neither is wrong. They simply feel differently; they are different persons.

The Thought then is the expression, the outgrowth, the covering, of underlying Feeling. And in the great life of Man as a whole, as in the lesser life of the individual, his continual new birth and inward growth causes his thought-systems also continually to change and be replaced by new ones. Like the bud-sheaths and husks in a growing plant or tree they give form for a time to the life within; then they fall off and are replaced. The husk prepares the bud underneath which is to throw it off. The thought prepares and protects the feeling underneath which growing will inevitably reject it; and when a thought has been formed it is already *false*, *i.e.*, ready to fall.

We are now, then, in a position to come back to the question of a genuine Science, truly so-called.

As there is no invariable and absolute datum on the fringe of Humanity—no definable flying atom on which we can found our reasonings—and as Modern Science, considered as an actual representation of the universe, falls miserably to pieces in consequence—is it possible that we have made a mistake in the *direction* in which we have sought for our datum; and may it be that we should look for that in the very Centre of Humanity instead of in its remotest circumference? In that direction evidently, if we could penetrate, we should expect to find, not a shadowy intellectual generalisation, but the very opposite of that—an intense immutable *feeling* or state, an axiomatic condition of Being. Is it possible that here, blazing like a sun (if we could only see it—and the sun is its allegory in the physical world), there exists within us absolutely such a thing—the one *fact* in the universe of which all else are shadows, *to* which everything has relation, and round which, itself unanalysable, all thought circles and all phenomena stand as indirect modes of expression?

Is it possible? That is the question—the question which each one of us has to solve. At any rate, let us throw this out as a suggestion. Let us suggest that as we have got nothing satisfactory by cleaning the sense-element out of phenomena, we should take the opposite course and put as much sense into them as we can!

"Facts" are, at least, half feelings. Let us acknowledge this and not empty the feeling out of them, but deepen and enlarge that which we already have in them. Who knows whether we have ever *seen* the blue sky? Who knows whether we have ever seen each other? Is it not a commonplace to say that one man sees in the common objects of Nature what another is wholly unconscious of? "The primrose on the river's brim a yellow primrose is to him—and nothing more." To what extent may the facts of Nature thus be deepened and made more substantial to us—and whither will this process lead us?

Do we not want to feel *more*, not less, in the presence of phenomena—to enter into a living relation with the blue sky, and the incense-laden air, and the plants and the animals—nay, even with poisonous and hurtful things to have a keener *sense* of their hurtfulness? Is it not a strange kind of science, that which wakes the mind to pursue the shadows of things, but dulls the senses to the reality of them—which causes a man to try to bottle the pure atmosphere of heaven and then to shut himself in a gas-reeking, ill-ventilated laboratory while he analyses it; or allows him to vivisect a dog, unconscious that he is blaspheming the pure and holy relation between man and the animals in doing so? Surely the man of Science (in its higher sense, that is) should be lynx-eyed as an Indian, keen-scented as a hound—with all senses and feelings trained by constant use and a pure and healthy life in close contact with Nature, and with a heart beating in sympathy with every creature. Such a man would have at command, so to speak, the key-board of the universe; but the mechanical, unhealthy, indoor-living student—is he not really *ignorant of the facts?*—Certainly, since he has not felt them, he is.

The process of the true Science consists first in the naming and defining of phenomena (*i.e.*, the facts of human consciousness), and secondly, in the discovery of the true relation of these phenomena to each other; and since the definitions of phenomena and their relations keep varying with the standpoint of the observer, the process evidently involves all experience, and ultimately the discovery of that last fact of experience to which and through which all the other facts are related. It is therefore an age-long process, and has to do with the emotional and moral part of man as well as with the logical and intellectual. It is in fact the discovery of the nature of Man himself, and of the true order of his being.

Modern Science—though seeking for a unity in Nature—fails to find it, because, from the nature of the case, any large body of knowledge in which all people will agree is limited to certain small regions of human experience—regions in which very likely no unity is discoverable. It takes the emerald, and breaks it up; treats of its color and light-refracting qualities on the one hand; of its crystalline structure and hardness on the other; of its weight and density; and of its chemical properties; all separately, and producing long strings of generalisation from each aspect of the subject. But how all these qualities are conjoined together, what their relation is which *constitutes* the emerald—yea, even the smallest bit of emerald dust—it (wisely) does not attempt to say. It takes the man and dissects him; treats of his blood, his nerves, his bones, his brain; of his senses of sight, of touch, of hearing; but of that which binds these together into a unity, of their true relation to each other in the man, it is silent.

Yet the man knows of himself that he *is* a unity; he knows

that all parts of his body have relation to *him*, and to each other; he knows that his senses of sight and hearing and touch and taste and smell are conjoined in the focus of his individual life, in his "I am;" he knows that all his faculties and powers, however much they may belong to different planes, spiritual or material, or may come under the inquisition of different Sciences, have an order of their own among each other—that there *is* an ultimate Science of them—even though he be not yet wholly versed in it. And he knows moreover that in a grain of dust, or in an emerald, or in an orange, or in any object of Nature, the different attributes of the object—which the Sciences thus treat of separately—are only the reflexion of his different senses; so that the problem of the conjunction of different attributes in a body comes back to the same problem of the union of various senses and powers in himself—each individual object being only a case, externalised as it were, and made a matter of consciousness, of the general relation to each other of his own sensations and feelings. Knowing all this—I say—he sees that the understanding of Nature in general and of the laws or relations which he thinks he perceives among external things, must always depend on the relations and laws which he tacitly assumes, or which he is directly conscious of, as existing between the various parts of his own being; and that the ultimate truth which Science—the divine Science—is really in search of is a moral Truth—an understanding of what man is, and the discovery of the true relation to each other of all his faculties—involving all experience, and an exercise of every faculty, physical, intellectual, emotional and spiritual, instead of one set of faculties only.

Not till we know the law of ourselves, in fact, shall we know the law of the emerald and the orange, or of Nature generally; and the law of ourselves is not learnt, except subordinately, by intellectual investigation; it is mainly learnt by life. The relation of gravity to vitality is learnt not so much by outer experiment in a laboratory as by long experience within ourselves from the day when as infants we cannot lift ourselves above the floor, through the years of the proud strength of manhood scaling the loftiest mountains, to the hour when our disengaged spirits finally overcome and pass beyond the attraction of the earth; and just as the sense of weight—which first appears as a quite external sensation—is thus at last found to stand in most pregnant relation with our deepest selves, so of the other senses which feed the individual life—the senses of light, of warmth, of taste, of sound, of smell. Taste, which begins as it were on the tip of the tongue, becomes ultimately, if normally developed, a sense which identifies itself with the health and well-being of the whole body; the pleasure of taste becomes vastly more than a mere surface pleasure, and its discrimination of food more than a mere regard for the nutrition of the ordinary corporeal functions. The sense of Light, which begins in the material eye, grows and deepens inwardly till the consciousness of it pervades the whole body and mind with a kind of inward illumination or divine Reason, showing the places of all things and enfolding the sense of beauty in itself. The sense of Warmth in the same manner is related to and leads up to Love; and Sound, in the voices of our friends or the divine chords of music, has passed away from being an external phenomenon and has established itself as the language of our most tender and intimate emotions.

All the senses thus as they develop and deepen are found to unite in the very focus of individual life. Slowly, and through long experience, their relation to each other, *their very meaning* unfolds, or will unfold; and as this process takes place the man knows himself *one*, a unity, of which the various faculties are the different manifestations. Then further through his less localised feelings or more glorified senses the individual finds his relation to other individuals. Through his loves and hatreds, through his senses of attraction, repulsion, cohesion, solidarity, order, justice, charity, right, wrong and the rest—these feelings, each like the others deepening back more and more as time goes on—he gradually discovers his true and abiding relationship to other individuals, and to the divine society of which they all form a part—and so at last, if we may venture to say so, his relationship to the absolute and universal. At present, since our most important relation to each other is conceived of as one of rivalry and Competition, we of course think of the objects of Nature as being chiefly engaged in a Struggle for Existence with each other; but when we become aware of all our senses and feelings, and of ourselves as individuals, as having relation to the Absolute and universal, proceeding from it, as the branches and twigs of a tree from the trunk—then we shall become aware of a Divine or absolute science in Nature; we shall at last understand that all objects have a permanent and indissoluble relation to each other, and shall see their true meaning—though not till then.

Is it possible then that Science, having hitherto—and we shall see in time that this process has been really most valuable and important—gone outwards from the centre towards the very fringe of Humanity—emptying facts as far as possible as it went of all feeling, and reducing itself at last to the most shadowy generalisations on the very verge of sense and nonsense—is it possible, I say, that it will now return, and *first* filling up facts with feeling as far as practicable (that is, by direct and the most living contact with Nature in every form, learning to enter into direct personal sense-relationship with every phenomenon and phase), will so gradually ascend to the great central fact and feeling, and then at last and for the first time become fully conscious of a vast organisation—absolutely perfect and intimately knit from its centre to its utmost circumference—(the true cosmos of Man—the conceptions of man and god combined)—existing inchoate or embryonic in every individual man, animal, plant, or other creature—the object of all life, experience, suffering, and toil—the ground of all sensation, and the hidden yet proper theme of all thought and study?

For this is it possible that Science will, speaking broadly, have to leave the laboratory and become one with Life; or that the great currents of human life will have to be turned on into these often Augean stables of intellectual pruriency? —the investigation of Nature no longer a matter of the intellect alone, but of patient listening and the quiet eye, and of love and faith, and of all deep human experience, bearing not superciliously its weight towards the interpretation of the least phenomenon—every "fact" thus deepened to its utmost—all experience (rather than experiment) courted, and filial walking with Nature, rather than tearing of veils aside—the life of the open air, and on the land and the waters, the companionship of the animals and the trees and the stars, the knowledge of their habits at first hand and through individual relationship to them, the recognition of their voices and languages, and listening well what they themselves have to say; the keenest education of the senses towards the physical powers and elements, and the acceptance of *all* human experience, without exception—till Science become a reality.

Is it possible that in some sense, instead of reducing each branch of Science to its lowest terms, we shall have to read it in the light of its *highest* factors, and "take it up" into the Scienceabove—that we shall have to take up the mechanical sciences into the physical, the physical into the vital, the vital into the social and ethical, and so forth, before we can understand them? Is it possible that the phenomena of Chemistry only find their due place and importance in their relation to living beings and processes; that the phenomena of Vitality and the laws of Biology and Zoology—Evolution included—can only be "explained" by their dependence on self-hood—both in plants and animals; that Political Economy and the Social Sciences (which deal with men as individual selves) must, to be understood aright, be studied in the light of those great ethical principles and enthusiasms, which to a certain extent override the individual self; and that, finally, Ethics or the study of moral problems is only comprehensible when the student has become aware of a region beyond Ethics, into which questions of morality and immorality, of right and wrong, do not and cannot enter?

Of this reversal of the ordinary scientific method Ruskin has given a great and signal instance in his treatment of Political Economy; it remains, perhaps, for others to follow his example in the other branches of Science.[1]

With regard to the absolute datum question we have seen that Science has two alternatives before it—either to be merely intellectual and to seek for its start-point in some quite external (and imaginary) thing like the Atom, or to be divine and to

[1] Thus the study of Geometry would be primarily an education of the eye, and the mind's eye, to the perception of geometrical forms and facts, the judgment of angles, &c.—and secondarily only a process of deductive reasoning—a body of empirical knowledge strengthened and tied together by bands of logic; the study of Natural History would be primarily an affectionate intimacy with the habits of animals and plants, and classification would be treated as a secondary matter and as a help to the former; Physiology would be studied in the first place by the method of Health—the pure body—becoming gradually transparent with all its organs to the eye of the mind—and dissection would be used to corroborate and correct the results thus attained; and so on.

seek for its absolute in the innermost recesses of humanity. We have two similar alternatives in the doctrine of Evolution, which looks either to one end of the scale or the other for its interpretation—either to the amoeba or to the man—to something it knows next to nothing of, or to that which it knows most of. Goethe, when gazing at a fan-palm at Padua, conceived the idea of leaf metamorphosis, which he afterwards ennunciated in the now accepted doctrine that all parts of a plant—seed-vessel, pistil, stamens, petals, sepals, stalk, &c.—may be regarded as modifications of a leaf or leaves. In this view the distinctions between the parts are effaced, and we have only one part instead of many—but the question is "what is that part?" It is of course arbitrary to call it a leaf, for since it is continually varying it is at one time a leaf, and at another a stalk, and then a petal or a sepal, and so forth. What then is it? For the moment we are baffled.

So with the doctrine of Evolution as applied to the whole organic kingdom up to man. Like the doctrine of leaf-metamorphosis it obliterates distinctions. Geoffroy St. Hilaire proposed to show the French Academy that a Cephalopod could be assimilated to a Vertebrate by supposing the latter bent backwards and walking on its hands and feet. There is a continuous variation from the mollusc to the man—all the lines of distinction run and waver—classes and species cease to exist—and Science instead of many sees only *one* thing. What then is that one thing? Is it a mollusc, or is it a man, or what is it? Are we to say that man may be looked upon as a variation of a mollusc or an amoeba, or that the amoeba may be looked on as a variation of man? Here are two directions of thought; which shall we choose? But the plain truth is, the Intellect can give no satisfactory answer. Whichever, or whatever, it chooses, the choice is quite arbitrary—just as much so as the choice of the "leaf" in the other case. There is no answer to be given. And thus it is that *the appearance of the doctrine of Evolution is the signal of the destruction of Science* (in the ordinary acceptation of the word). For Evolution is the successive obliteration of the arbitrary distinctions and landmarks which by their existence *constitute* Science, and as soon as Evolution covers the whole ground of Nature—inorganic and organic (as before long it will do)—the whole of Nature runs and wavers before the eye of Science, the latter recognises that its distinctions *are* arbitrary, and turns upon and destroys itself. This has happened before, I believe—ages back in the history of the human race—and probably will happen again.

The only conceivable answer to the question, "What is that which is now a mollusc and now a man and now an inorganic atom?"[1] is given by man himself—and his answer is, I fear, not "scientific." It is "I Am." "I am that which varies." And the force of his answer depends on what he means by the word "I." And so also the only conceivable answer to the absolute datum question is to be found in the meaning of the word "I"—in the deepening back of consciousness itself. Man is the measure of all things. If we are to use Science as a minister to the most external part of man—to provide him with cheap boots and shoes, &c.—then we do right to seek our absolute datum in his external part, and to take his *foot* as our first measure. We found a science on feet and pounds, and it serves its purpose well enough. But if we want to find a garment for his inner being—or, rather, one that shall fit the *whole* man—to wear which will be a delight to him and as it were a very interpretation of himself—it seems obvious that we must not take our measure from outside, but from his very most central principle. The whole question is, whether

[1] Compare the Sphinx-riddle: What is that which goes on four legs, &c.

there *is* any absolute datum in this direction or not. There have been men through all ages of history (and from before) who have declared that there is. They have perhaps been conscious of it in themselves. On the other hand there have been men who, starting from their feet, declared that consciousness itself was a mere incident of the human machine—as the whistle of the engine—and thus the matter stands. On the whole, at the present day, the *feet* have it, and (notwithstanding their variety in size and boot-induced conformation) are generally accepted as the best absolute datum available.

Under the foot *régime* the universe is generally conceived of as a medley of objects and forces, more or less orderly and distinct from man, in the midst of which man is placed—the purpose and tendency of his life being "adaptation to his environment." To understand this we may imagine Mrs. Brown in the middle of Oxford Street. 'Buses and cabs are running in different directions, carts and drays are rattling on all sides of her. This is her environment, and she has to adapt herself to it. She has to learn the laws of the vehicles and their movements, to stand on this side or on that, to run here and stop there, conceivably to jump into one at a favorable moment, to make use of the law of its movement, and so get carried to her destination as comfortably as may be. A long course of this sort of thing "adapts" Mrs. Brown considerably, and she becomes more active, both in mind and body, than before. That is all very well. But Mrs. Brown has a *destination*. (Indeed how would she ever have got into the middle of Oxford Street at all if she had not had one? and if she did get there with no destination at all, but merely to skip about, would there be any Mrs. Brown left in a short time?) The question is, "What is the destination of Man?"

About this last question unfortunately we hear little. The theory is (I hope I am not doing it injustice) that by studying your environment sufficiently you will find out—that is, that by investigating Astronomy, Biology, Physics, Ethics, &c., you will discover the destiny of man. But this seems to me the same as saying that by studying the laws of cabs and 'buses sufficiently you will find out where you are going to. These are ways and means. Study them by all means, that is right enough; but do not think *they* will tell you where to go. You have to use them, not they you.

In order therefore for the environment to act, there must be a destination. This I suppose is expressed in the biological dictum, "organism is made by function as well as environment." What then is the function of Man? And here we come back again to the meaning of the word "I."

Notwithstanding then the prevalence of the foot régime, and that the heathen so furiously rage together in their belief in it, let us suggest that there is in man a divine consciousness as well as a foot-consciousness. For as we saw that the sense of taste may pass from being a mere local thing on the tip of the tongue to pervading and becoming synonymous with the health of the whole body; or as the blue of the sky may be to one person a mere superficial impression of color, and to another the inspiration of a poem or picture, and to a third—as to the "god-intoxicated" Arab of the desert—a living presence like the ancient Dyaus or Zeus; so may not the whole of human consciousness gradually lift itself from a mere local and temporary consciousness to a divine and universal? There is in every man a local consciousness connected with his quite external body; that we know. Are there not also in every man the makings of a universal consciousness? That there are in us phases of consciousness which transcend the limit of the bodily senses is a matter of daily experience; that we perceive and know things which are not conveyed to us by our bodily eyes or heard by our bodily ears is certain; that there rise in us waves of consciousness from those around us, from the people, the race, to which we belong, is also certain; may there not then be in us the makings of a perception and knowledge which shall not be relative to this body which is here and now, but which shall be good for all time and everywhere? Does there not exist in truth as we have already hinted—an inner Illumination—of which what we call light in the outer world is the partial expression and manifestation—by which we can ultimately see things *as they are*, beholding all creation, the animals, the angels, the plants, the figures of our friends and all the ranks and races of human kind, in their true being and order—not by any local act of perception but by a cosmical intuition and presence, identifying ourselves with what we see? Does there not exist a perfected sense of Hearing—as of the morning-stars singing together—an understanding of the words that are spoken all through the universe, the hidden meaning of all things, the word which is creation itself—a profound and far pervading sense, of which our ordinary sense of sound is only the first novitiate and initiation? Do we not become aware of an inner sense of Health and of holiness—the translation and final outcome of the external sense of taste—which has power to determine for us absolutely and without any ado, without argument and without denial, what is good and appropriate to be done or suffered in every case that can arise?

And so on; it is not necessary to say more. If there are such powers in man, then there is indeed an exact science possible. Short of it there is only a temporary and phantom science. "Whatever is known to us by (direct) consciousness," says Stuart Mill in his System of Logic, "is known to us beyond possibility of question;" what is known by our local and temporary consciousness is known *for the moment* beyond possibility of question; what is known by our permanent and universal consciousness is permanently known beyond possibility of question.

DEFENCE OF CRIMINALS:

A CRITICISM OF MORALITY.

The State is the actually existing realised moral life. For it is the unity of the universal essential Will with that of the individual, and this is "Morality."—Hegel.

A criminal is literally a person accused—accused, and in the modern sense of the word convicted, of being harmful to Society. But is he there in the dock, the patch-coated brawler or burglar, really harmful to Society? is he more harmful than the mild old gentleman in the wig who pronounces sentence upon him? That is the question. Certainly he has infringed the law: and the law is in a sense the consolidated public opinion of Society: but if no one were to break the law, public opinion would ossify, and society would die. As a matter of fact Society keeps changing its opinion. How then are we to know when it is right and when it is wrong? The Outcast of one age is the Hero of another. In execration they nailed Roger Bacon's manuscripts out in the sun and rain, to rot crucified upon planks—his bones lie in an unknown and unhonored grave—yet to-day he is regarded as a pioneer of human thought. The hated Christian holding his ill-famed love-feasts in the darkness of the catacombs has climbed on to the throne of S. Peter and the world. The Jew money-lender whom Front-de-Bœuf could torture with impunity is become a Rothschild—guest of princes and instigator of commercial wars; and

Shylock is now a highly respectable Railway Bondholder. And the Accepted of one age is the Criminal of the next. All the glories of Alexander do not condone in our eyes for his cruelty in crucifying the brave defenders of Tyre by thousands along the sea-shore; and if Solomon with his thousand wives and concubines were to appear in London to-morrow, even our most frivolous circles would be shocked, and Brigham Young by contrast seem a domestic model. The judge pronounces sentence on the prisoner now, but Society in its turn and in the lapse of years pronounces sentence on the judge. It holds in its hand a new canon, a new code of morals, and consigns its former representative and the law which he administered to a limbo of contempt.

It seems as if Society, as it progresses from point to point, forms ideals—just as the individual does. At any moment each person, consciously or unconsciously, has an ideal in his mind toward which he is working (hence the importance of literature). Similarly Society has an ideal in its mind. These ideals are tangents or vanishing points of the direction in which Society is moving at the time. It does not reach its ideal, but it goes in that direction—then, after a time, the direction of its movement changes, and it has a new ideal.

When the ideal of Society is material gain or possession, as it is largely to-day, the object of its special condemnation is the thief—not the rich thief, for he is already in possession and therefore respectable, but the poor thief. There is nothing to show that the poor thief is really more immoral or unsocial than the respectable money-grubber; but it is very clear that the money-grubber has been floating with the great current of Society, while the poor man has been swimming against it, and so has been worsted. Or when, as to-day, Society rests on private property in land, its counter-ideal is the poacher. If you go in the company of the county squire-archy and listen to the after-dinner talk you will soon think the poacher a combination of all human and diabolic vices; yet I have known a good many poachers, and either have been very lucky in my specimens or singularly prejudiced in their favor, for I have generally found them very good fellows—but with just this one blemish that they invariably regard a landlord as an emissary of the evil one! The poacher is as much in the right, probably, as the landlord, but he is not right for the time. He is asserting a right (and an instinct) belonging to a past time—when for hunting purposes all land was held in common—or to a time in the future when such or similar rights shall be restored. Cæsar says of the Suevi that they tilled the ground in common, and had no private lands, and there is abundant evidence that all early human communities before they entered on the stage of modern civilisation were communistic in character. Some of the Pacific Islanders to-day are in the same condition. In those times private property was theft. Obviously the man who attempted to retain for himself land or goods, or who fenced off a portion of the common ground and—like the modern landlord—would allow no one to till it who did not pay him a tax—was a criminal of the deepest dye. Nevertheless the criminals pushed their way to the front, and have become the respectables of modern Society. And it is quite probable that in like manner the criminals of to-day will push to the front and become the respectables of a later age.

The ascetic and monastic ideal of early Christian and mediæval ages is now regarded as foolish, if not wicked; and poverty, which in many times and places has been held in honor as the only garb of honesty, is condemned as criminal and indecent. Nomadism—if accompanied by poverty—is criminal in modern Society. To-day the gipsy and the tramp are hunted down. To have no settled habitation, or worse still, no place to lay your head, are suspicious matters. We close even our outhouses and barns against the son of man, and so to us the son of man comes not. And yet—at one time and in one stage of human progress—the nomadic state is the rule; and the settler is then the criminal. His crops are fired and his cattle driven off. What right has he to lay a limit to the hunting grounds, or to spoil the wild free life of the plains with his dirty agriculture?

As to the marriage relation and its attendant moralities, the forms are numerous and notorious enough. Public opinion seems to have varied through all phases and ideals, and yet there is no indication of finality. Modern investigations show that in primitive human societies the affinities admitted or barred in marriage are most various—the relation of brother and sister being even in cases allowed; in the present day such a bond as the last-mentioned would be considered inhuman and monstrous.[1] Polyandry prevails among one people or at one time, polygyny prevails among another people or at another time. In Central Africa to-day the chief offers you his wife as a mark of hospitality, in India the native Prince keeps her hidden even from his most intimate guest. Among the Japanese, public opinion holds young women—even of good birth—singularly free in their intercourse with men, *till they are married;* at Paris they are free after. In the Greek and Roman antiquity marriage seems with some brilliant exceptions to have been a prosaic affair—mostly a matter of convenience and housekeeping—the woman an underling—little of the ideal attaching to the relationship of

[1] Yet there is no doubt that lasting and passionate love may exist between two persons thus nearly related. The danger to the health of the offspring from occasional in-breeding of the kind appears to arise chiefly from the accentuation of infirmities common to the two parents. In a state of society free from the diseases of the civilisation-period, such a danger would be greatly reduced.

man and wife. The romance of love went elsewhere. The better class of free women or Hetairai were those who gave a spiritual charm to the passion. They were an educated and recognised body, and possibly in their best times exercised a healthy and discriminating influence upon the male youth. The respectful treatment of Theodota by Socrates, and the advice which he gives her concerning her lovers: to keep the insolent from her door, and to rejoice greatly when the accepted succeed in anything honorable, indicates this. That their influence was at times immense the mere name of Aspasia is sufficient to show; and if Plato in the Symposium reports correctly the words of Diotima, her teaching on the subject of human and divine love was probably of the noblest and profoundest that has ever been given to the world.

With the influx of the North-men over Europe came a new ideal of the sexual relation, and the wife mounted more into equality with her husband than before. The romance of love, however, still went mainly outside marriage, and may I believe be traced in two chief forms—that of Chivalry, as an ideal devotion to pure Womanhood; and that of Minstrelsy, which took quite a different hue, individual and sentimental—the lover and his mistress (she in most cases the wife of another), the serenade, secret amour, &c.—both of which forms of Chivalry and Minstrelsy contain in themselves something new and not quite familiar to antiquity.

Finally in modern times the monogamic union has risen to pre-eminence—the splendid ideal of an equal and life-long attachment between man and wife, fruitful of children in this life, and hopeful of continuance beyond—and has become the great theme of romantic literature, and the climax of a thousand novels and poems. Yet it is just here and to-day, when this ideal after centuries of struggle has established itself, and among the nations that are in the van of civilisation—that we find the doctrine of perfect liberty in the marriage relationship being most successfully preached, and that the communalization of social life in the future seems likely to weaken the family bond and to relax the obligation of the marriage tie.

If the Greek age, splendid as it was in itself and in its fruits to human progress, did not hold marriage very high, it was partly because the ideal passion of that period, and one which more than all else inspired it, was that of comradeship, or male friendship carried over into the region of love. The two figures of Harmodius and Aristogiton stand at the entrance of Greek history as the type of this passion, bearing its fruit (as Plato throughout maintains is its nature) in united self-devotion to the country's good. The heroic Theban legion, the "sacred band," into which no man might enter without his lover—and which was said to have remained unvanquished till it was annihilated at the battle of Chæronea—proves to us how publicly this passion and its place in society were recognised; while its universality and the depth to which it had stirred the Greek mind are indicated by the fact that whole treatises on love, in its spiritual aspect, exist, in which no other form of the sentiment seems to be contemplated; and by the magnificent panorama of Greek statuary, which was obviously to a large extent inspired by it. In fact the most remarkable Society known to history, and its greatest men, can not be properly considered or understood apart from this passion; yet the modern world scarcely recognises it, or if it recognises, does so chiefly to condemn it.[1]

[1] Modern writers fixing their regard on the physical side of this love (necessary no doubt here, as elsewhere, to define and corroborate the spiritual) have entered their protest as against the mere obscenity into which the thing fell—for instance in the days of Martial—but have missed the profound significance of the heroic attachment itself. It is, however, with the ideals that we are just now concerned and not with their disintegration.

Other instances might be quoted to show how differently moral questions are regarded in one age and another—as in the case of Usury, Magic, Suicide, Infanticide, &c. On the whole we pride ourselves (and justly I believe) on the general advance in humanity; yet we know that to-day the merest savages can only shudder at a civilisation whose public opinion allows—as amongst us—the rich to wallow in their wealth while the poor are systematically starving; and it is certain that the vivisection of animals—which on the whole is approved by our educated classes (though not by the healthier sentiment of the uneducated)—would have been stigmatised as one of the most abominable crimes by the ancient Egyptians—if, that is, they could have conceived such a practice possible at all.

But not only do the moral judgments of mankind thus vary from age to age and from race to race, but—what is equally remarkable—they vary to an extraordinary degree from class to class of the same society. If the landlord class regards the poacher as a criminal, the poacher as already hinted looks upon the landlord as a selfish ruffian who has the police on his side; if the respectable shareholder, politely and respectably subsisting on dividends, dismisses navvies and the frequenters of public-houses as disorderly persons; the navvy in return despises the shareholder as a sneaking thief. And it is not easy to see, after all, which is in the right. It is useless to dismiss these discrepancies by supposing that one class in the nation possesses a monopoly of morality and that the other classes simply rail at the virtue they cannot attain to, for this is obviously not the case. It is almost a commonplace, and certainly a fact that cannot be contested, that every class—however sinful or outcast in the eyes of others—contains within its ranks a large proportion of generous, noble, self-sacrificing characters; so that the public opinion of one such class, however different from that of others, cannot at least be invalidated on the above ground. There are plenty of clergymen at this moment who are models of pastors—true shepherds of the people—though a large and increasing section of society persist in regarding priests as a kind of wolves in sheep's clothing. It is not uncommon to meet with professional thieves who are generous and open-handed to the last degree, and ready to part with their last penny to help a comrade in distress; with women living outside the bounds of conventional morality who are strongly religious in sentiment, and who regard atheists as *really* wicked people; with aristocrats who have as stern material in them as quarry-men; and even with bondholders and drawing-room loungers who are as capable of bravery and self-sacrifice as many a pitman or ironworker. Yet all these classes mentioned have their codes of morality, differing in greater or lesser degree from each other; and again the question forces itself upon us: Which of them all is the true and abiding code?

It may be said, with regard to this variation of codes within the same society, that though various codes may exist at the same time, one only is really valid, namely that which has embodied itself in the law—that the others have been rejected because they were unworthy. But when we come to look into this matter of law we see that the plea can hardly be maintained. Law represents from age to age the code of the dominant or ruling class, slowly accumulated, no doubt, and slowly modified, but always added to and always administered by the ruling class. To-day the code of the dominant class may perhaps best be denoted by the word Respectability—and if we ask why this code has to a great extent overwhelmed the codes of the other classes and got the law on its side (so far that in the main it characterises those classes who do not conform to it as the criminal classes), the answer can only be: Because it *is* the code of the classes who are in power.

Respectability is the code of those who have the wealth and the command, and as these have also the fluent pens and tongues, it is the standard of modern literature and the press. It is not necessarily a better standard than others, but it is the one that happens to be in the ascendant; it is the code of the classes that chiefly represent modern society; it is the code of the Bourgeoisie. It is different from the Feudal code of the past, of the knightly classes, and of Chivalry; it is different from the Democratic code of the future—of brotherhood and of equality; it is the code of the Commercial age—and its distinctive watchword is—property.

The respectability of to-day is the respectability of property. There is nothing so respectable as being well-off. The Law confirms this: everything is on the side of the rich; justice is too expensive a thing for the poor man. Offences against the person hardly count for so much as those against property. You may beat your wife within an inch of her life and only get three months; but if you steal a rabbit, you may be "sent" for years. So again gambling by thousands on Change is respectable enough, but pitch and toss for halfpence in the streets is low, and must be dealt with by the police; while it is a mere commonplace to say that the high-class swindler is "received" in society from which a more honest but patch-coated brother would infallibly be rejected. As Walt Whitman has it "There is plenty of glamour about the most damnable crimes and hoggish meannesses, special and general, of the feudal and dynastic world over there, with its personnel of lords and queens and courts, so well-dressed and handsome. But the people are ungrammatical, untidy, and their sins gaunt and ill-bred."

Thus we see that though there are for instance in the England of to-day a variety of classes, and a variety of corresponding codes of public opinion and morality, one of these codes, namely that of the ruling class whose watchword is property, is strongly in the ascendant. And we may fairly suppose that in any nation from the time when it first becomes divided into well-marked classes this is or has been the case. In one age—the commercial age—the code of the commercial or money-loving class is dominant; in another—the military—the code of the warrior class is dominant; in another—the religious—the code of the priestly class; and so on. And even before any question of division into classes arises, while races are yet in a rudimentary and tribal state, the utmost diversity of custom and public opinion marks the one from the other.

What, then, are we to conclude from all these variations (and the far greater number which I have not mentioned) of the respect or stigma attaching to the *same* actions, not only among different societies in different ages or parts of the world, but even at any one time among different classes of the same society? Must we conclude that there is no such thing as a permanent moral code valid for all time; or must we still suppose that there is such a thing—though society has hitherto sought for it in vain?

I think it is obvious that there is no such thing as a permanent moral code—at any rate as applying to *actions*. Probably the respect or stigma attaching to particular classes of actions arose from the fact that these classes of actions were—or were thought to be—beneficial or injurious to the society of the time; but it is also clear that this good or bad name once created clings to the action long after the action has ceased in the course of social progress to be beneficial in the one case, or injurious in the other; and indeed long after the thinkers of the race have discovered the discrepancy. And so in a short time arises a great confusion in the popular mind between what is really good or evil for the race and what is reputed to be so—the bolder spirits who try to separate the two having to atone for this confusion by their own martyrdom. It is also pretty clear that the actions which are beneficial or injurious to the race must by the nature of the case vary almost indefinitely with the changing conditions of the life of the race—what is beneficial in one age or under one set of conditions being injurious in another age or under other circumstances—so that a permanent or ever-valid code of moral action is not a thing to be expected, at any rate by those who regard morality as a result of social experience, and as a matter of fact is not a thing that we find existing. And, indeed, of those who regard morals as intuitive, there are few who have thought about the matter who would be inclined to say that any *act* in itself can be either right or wrong. Though there is a superficial judgment of this kind, yet when the matter comes to be looked into, the more general consent seems to be that the rightness or wrongness is in the *motive.* To kill (it is said) is not wrong, but to do so with murderous intent is; to take money out of another person's purse is in itself neither moral nor immoral—all depends upon whether permission has been given, or on what the relations between the two persons are; and so on. Obviously there is no mere act which under given conditions may not be justified, and equally obviously there is no mere act which under given conditions may not become unjustifiable. To talk, therefore, about virtues and vices as permanent and distinct classes of actions is illusory: there is no such distinction, except so far as a superficial and transient public opinion creates it. The theatre of morality is in the passions, and there are (it is said) virtuous and vicious passions—eternally distinct from each other.

Here, then, we have abandoned the search for a permanent moral code among the actions; on the understanding that we are more likely to find such a thing among the passions. And I think it would be generally admitted that this is a move in the right direction. There are difficulties however here, and the matter is not one which renders itself up at once. Though, vaguely speaking, some passions seem nobler and more dignified than others, we find it very difficult, in fact impossible, to draw any strict line which shall separate one class, the virtuous, from the other class, the vicious. On the whole we place Prudence, Generosity, Chastity, Reverence, Courage, among the virtues—and their opposites, as Rashness, Miserliness, Incontinence, Arrogance, Timidity, among the vices; yet we do not seem able to say that Prudence is always better than Rashness, Chastity than Incontinence, or Reverence than Arrogance. There are situations in which the less honored quality is the most in place; and if the extreme of this is undesirable, the extreme of its opposite is undesirable too. Courage, it is commonly said, must not be carried over into foolhardiness; Chastity must not go so far as the monks of the early Church took it; there is a limit to the indulgence of the instinct of Reverence. In fact the less dignified passions are necessary sometimes as a counterbalance and set-off to the more dignified, and a character devoid of them would be very insipid; just as among the members of the body, the less honored have their place as well as the more honored, and could not well be discarded.

Hence a number of writers, abandoning the attempt to draw a fixed line between virtuous and vicious passions, have boldly maintained that vices have their place as well as virtues, and that the true salvation lies in the golden mean. The ἐπιείκεια and σωφροσύνη of the Greeks seems to have pointed to the idea of a blend or harmonious adjustment of all the powers as the perfection of character. Plutarch says (*Essay on Moral Virtue*), "This, then, is the function of practical reason following nature, to prevent our passions either going too far or too short. . . . Thus setting bound to the emotional currents, it creates in the unreasoning part of the soul moral habits which are the mean between excess and deficiency."

The English word "gentleman" seems to have once conveyed a similar idea. And Emerson, among others, maintains that each vice is only the "excess or acridity of a virtue," and says "the first lesson of history is the good of evil."

According to this view rightness or wrongness cannot be predicated of the passions themselves, but should rather be applied to the use of them, and to the way they are proportioned to each other and to circumstances. As, farther back, we left the region of actions to look for morality in the passions that lie behind action, so now we leave the region of the passions to look for it in the power that lies behind the passions and gives them their place. This is a farther move in the same direction as before, and possibly will bring us to a more satisfactory conclusion. There are still difficulties, however—the chief ones lying in the want of definiteness which necessarily attaches to our dealings with these remoter tracts of human nature; and in our own defective knowledge of these tracts.

For these reasons, and as the subject is a complex and difficult one, I would ask the reader to dwell for a few minutes longer on the considerations which show that it is really as impossible to draw a fixed line between moral and immoral passions as it is between moral and immoral actions, and which therefore force us if we are to find any ground of morality at all, to look for it in some further region of our nature.

Plato in his allegory of the soul—in the Phædrus—though he apparently divides the passions which draw the human chariot into two classes, the heavenward and the earthward—figured by the white horse and the black horse respectively—does not recommend that the black horse should be destroyed or dismissed, but only that he (as well as the white horse) should be kept under due control by the charioteer. By which he seems to intend that there is a power in man which stands above and behind the passions, and under whose control alone the human being can safely move. In fact if the fiercer and so-called more earthly passions were removed, half the driving force would be gone from the chariot of the human soul. Hatred may be devilish at times—but after all the true value of it depends on what you hate, on the use to which the passion is put. Anger though inhuman at one time is magnificent at another. Obstinacy may be out of place in a drawing-room, but it is the latest virtue on a battle-field when an important position has to be held against the full brunt of the enemy. And Lust, though maniacal and monstrous in its aberrations, cannot in the last resort be separated from its divine companion, Love. To let the more amiable passions have entire sway notoriously does not do: to turn your cheek, too literally, to the smiter, is (*pace* Tolstoi) only to encourage smiting; and when society becomes so altruistic that everybody runs to fetch the coal-scuttle we feel sure that something has gone wrong. The white-washed heroes of our biographies with their many virtues and no faults do not please us. We have an impression that the man without faults is, to say the least, a vague, uninteresting being—a picture without light and shade—and the conventional semi-pious classification of character into good and bad qualities (as if the good might be kept and the bad thrown away) seems both inadequate and false.

What the student of human nature rather has to do is not to divide the virtues (so-called) from the vices (so-called), not to separate the black horse and the white horse, but to find out what is the relation of the one to the other—to see the character as a whole, and the mutual interdependence of its different parts—to find out what that power is which constitutes it a unity, whose presence and control makes the man and all his actions "right," and in whose absence (if it is really possible for it to be entirely absent) the man and his actions must be "wrong."

What we call vices, faults, defects, appear often as a kind of limitation: cruelty, for instance, as a limitation of human sympathy, prejudice as a blindness, a want of discernment; but it is just these limitations—in one form or another—which are the necessary conditions of the appearance of a human being in the world. If we are to act or live at all we must act and live under limits. There must be channels along which the stream is forced to run, else it will spread and lose itself aimlessly in all directions—and turn no mill-wheels. One man is disagreeable and unconciliatory—the directions in which his sympathy goes out to others are few and limited—yet there are situations in life (and everyone must know them) when a man who is *able and willing* to make himself disagreeable is invaluable: when a Carlyle is worth any number of Balaams.

Sometimes again vices, &c., appear as a kind of raw material from which the other qualities have to be formed, and without which, in a sense, they could not exist. Sensuality, for instance, underlies all art and the higher emotions. Timidity is the defect of the sensitive imaginative temperament. Bluntness, stupid candor, and want of tact are indispensable in the formation of certain types of Reformers. But what would you have? Would you have a rabbit with the horns of a cow, or a donkey with the disposition of a spaniel? The reformer has not to extirpate his brusqueness and aggressiveness, but to see that he makes good use of these qualities; and the man has not to abolish his sensuality, but to humanise it.

And so on. Lecky, in his "History of Morals," shows how in society certain defects necessarily accompany certain excellencies of character. "Had the Irish peasants been less chaste they would have been more prosperous," is his blunt assertion, which he supports by the contention that their early marriages (which render the said virtue possible) "are the most conspicuous proofs of the national improvidence, and one of the most fatal obstacles to industrial prosperity." Similarly he says that the gambling table fosters a moral nerve and calmness "scarcely exhibited in equal perfection in any other sphere"—a fact which Bret Harte has finely illustrated in his character of Mr. John Oakhurst in the "Outcasts of Poker Flat;" also that "the promotion of industrial veracity is probably the single form in which the growth of manufactures exercises a favorable influence upon morals;" while, on the other hand, "Trust in Providence, content and resignation in extreme poverty and suffering, the most genuine amiability, and the most sincere readiness to assist their brethren, an adherence to their religious opinions which no persecutions and no bribes can shake, a capacity for heroic, transcendent, and prolonged self-sacrifice, may be found in some nations, in men who are habitual liars and habitual cheats." Again he points out that thriftiness and forethought—which, in an industrial civilisation like ours, are looked upon as duties "of the very highest order"—have at other times (when the teaching was "take no thought for the morrow") been regarded as quite the reverse, and concludes with the general remark that as society advances there is some loss for every gain that is made, and with the special indictment against "civilisation" that it is not favorable to the production of "self-sacrifice, enthusiasm, reverence, or chastity."

The point of all which is that the so-called vices and defects—whether we regard them as limitations or whether we regard them as raw materials of character, whether we regard them in the individual solely or whether we regard them in their relation to society—are necessary elements of human life, elements without which the so-called virtues could not exist; and that therefore it is quite impossible to separate vices and virtues into distinct classes with the latent idea involved that one class may be retained and the other in course of time got rid of. Defects and bad qualities will not be treated so—they clamor for their rights and will not be denied; they effect a lodgment in us, and we have to put up with them. Like the grain of sand in the oyster, we are forced to make pearls of them.

These are the precipices and chasms which give form to the mountain. Who wants a mountain sprawling indifferently out on all sides, without angle or break, like the oceanic tide-wave of which one cannot say whether it is a hill or a plain? And if you want to grow a lily, chastely white and filling the air with its fragrance, will you not bury the bulb of it deep in the dirt to begin with?

Acknowledging, then, that it is impossible to hold permanently to any line of distinction between good and bad passions, there remains nothing for it but to accept both, and to *make use* of them—redeeming them, both good and bad, from their narrowness and limitation by so doing—to make use of them in the service of humanity. For as dirt is only matter in the wrong place, so evil in man consists only in actions or passions which are uncontrolled by the human within him, and undedicated to its service. The evil consists not in the actions or passions themselves, but in the fact that they are inhumanly used. The most unblemished virtue erected into a barrier between oneself and a suffering brother or sister—the whitest marble image, howsoever lovely, set up in the Holy Place of the temple of Man, where the spirit alone should dwell—becomes blasphemy and a pollution.

Wherein exactly this human service consists is another question. It may be, and, as the reader would gather, probably is, a matter which at the last eludes definition. But though it may elude exact statement, that is no reason why approximations should not be made to the statement of it; nor is its ultimate elusiveness of intellectual definition any proof that it may not become a real and vital force within the man, and underlying inspiration of his actions. To take the two considerations in order. In the first place, as we saw from the beginning, the experience of society is continually leading it to classify actions into beneficial and harmful, good and bad; and thus moral codes are formed which eat their way from the outside into the individual man and become part of him. These codes may be looked upon as approximations in each age to a statement of human service; but, as we have seen, they are by the nature of the case very imperfect; and since the very conditions of the problem are continually changing, it seems obvious that a final and absolute solution of it by this method is impossible. The second way in which man works towards a solution is by the expansion and growth of his own consciousness, and is ultimately by far the most important—though the two methods have doubtless continually to be corrected by each other. In fact, as man actually forms a part of society externally, so he comes to know and *feel* himself a part of society through his inner nature. Gradually, and in the lapse of ages, through the development of his sympathetic relation with his fellows, the individual man enters into a wider and wider circle of life—the joys and sorrows, the experiences, of his fellows become his own joys and sorrows, his own experiences—he passes into a life which is larger than his own individual life—forces flow in upon him which determine his actions, not for results which return to him directly, but for results which can only return to him indirectly and through others; at last the ground of humanity, as it were, reveals itself within him, the region of human equality—and his actions come to flow directly from the very same source which regulates and inspires the whole movement of society. At this point the problem is solved. The growth has taken place from within; it is not of the nature of an external compulsion, but of an inward compunction. By actual consciousness the man has taken on an ever-enlarging life, and at last the life of humanity, which has no fixed form, no ever-valid code; but is itself the true life, surpassing definition, yet inspiring all actions and passions, all codes and forms, and determining at last their place.

It is the gradual growth of this supreme life in each individual which is the great and indeed the only hope of Society—it is that for which Society exists: a life which so far from dwarfing individuality enhances immensely its power, causing the individual to move with the weight of the universe behind him—and exalting what were once his little peculiarities and defects into the splendid manifestations of his humanity.

To return then for a moment to the practical bearing of this on the question before us, we see that so soon as we have abandoned all codes of morals there remains nothing for us but to put *all* our qualities and defects to human use, and to redeem them so by doing. Our defects are our entrances into life, and the gateway of all our dealings with others. Think what it is to be plain and *homely*. The very word suggests an endearment, and a liberty of access denied to the faultlessly handsome. Our very evil passions, so called, are not things to be ashamed of, but things to look straight in the face and to see what they are good for—for a use can be found for them, that is certain. The man should see that he is worthy of his passion, as the mountain should rear its crest conformable to the height of the precipice which bounds it. Is it women? let him see that he is a magnanimous lover. Is it ambition? let him take care that it be a grand one. Is it laziness? let it redeem him from the folly of unrest, to become heaven-reflect-ing, like a lake among the hills. Is it closefistedness? let it become the nurse of a true economy.

The more complicated, pronounced, or awkward the defect is the finer will be the result when it has been thoroughly worked up. Love of approbation is difficult to deal with. Through sloughs of duplicity, of concealment, of vanity, it leads its victim. It sucks his sturdy self-life, and leaves him flattened and bloodless. Yet once mastered, once fairly torn out, cudgeled, and left bleeding on the road (for this probably has to be done with every vice or virtue some time or other), it will rise up and follow you, carrying a magic key round its neck, meek and serviceable now, instead of dangerous and demoniac as before.

Deceit is difficult to deal with. In some sense it is the worst fault that can be. It seems to disorganise and ultimately to destroy the character. Yet I am bold to say that this defect has its uses. Severely examined perhaps it will be found that no one can live a day free from it. And beyond that—is not "a noble dissimulation" part and parcel of the very greatest characters: like Socrates, "the white soul in a satyr form?" When the divine has descended among men has it not always like Moses worn a veil before its face? and what is Nature herself but one long and organised system of deception?

Veracity has an opposite effect. It knits all the elements of a man's character—rendering him solid rather than fluid; yet carried out too literally and pragmatically it condenses and solidifies the character overmuch, making the man woodeny and angular. And even of that essential Truth (truth to the inward and ideal perfection) which more than anything else perhaps *constitutes* a man—it is to be remembered that even here there must be a limitation. No man can in act or externally be quite true to the ideal—though in spirit he may be. If he is to live in this world and be mortal, it must be by virtue of some partiality, some defect.

And so again—since there is an analogy between the Individual and Society—may we not conclude that as the individual has ultimately to recognise his so-called evil passions and find a place and a use for them, society also has to recognise its so-called criminals and discern their place and use? The artist does not omit shadows from his canvas; and the wise statesman will not try to abolish the criminal from society—lest haply he be found to have abolished the driving force from his social machine.[1]

From what has now been said it is quite clear that in general we call a man a criminal, not because he violates any eternal code of morality—for there exists no such thing—but because he violates the ruling code of his time, and this depends largely on the ideal of the time. The Spartans appear to have permitted theft because they thought that thieving habits in the community fostered military dexterity and discouraged the accumulation of private wealth. They looked upon the latter as a great evil. But to-day the accumulation of private wealth is our great good and the thief is looked upon as the evil. When however we find, as the historians of to-day teach us, that society is now probably passing through a parenthetical stage of private property from a stage of communism in the past to a stage of more highly developed communism in the future, it becomes clear that the thief (and the poacher beforementioned) is that person who is protesting against the too-exclusive domination of a passing ideal. Whatever should we do without him? He is keeping open for us, as Hinton I think expresses it, the path to a regenerate society, and is more use-

[1] The derivation of the word "wicked" seems uncertain. May it be suggested that it is connected with "wick" or "quick" meaning *alive*?

ful to that end than many a platform orator. He it is that makes Care to sit upon the Crupper of Wealth, and so, in course of time, causes the burden and bother of private property to become so intolerable that society gladly casts it down on common ground. Vast as is the machinery of Law, and multifarious the ways in which it seeks to crush the thief, it has signally failed, and fails ever more and more. The thief will win. He will get what he wants, but (as usual in human life !) in a way and in a form very different from what he expected.

And when we regard the thief in himself, we cannot say that we find him less human than other classes of society. The sentiment of large bodies of thieves is highly communistic among themselves; and if they thus represent a survival from an earlier age, they might also be looked upon as the precursors of a better age in the future. They have their pals in every town, with runs and refuges always open, and are lavish and generous to a degree to their own kind. And if they look upon the rich as their natural enemies and fair prey, a view which it might be difficult to gainsay, many of them at any rate are animated by a good deal of the Robin Hood spirit, and are really helpful to the poor.

I need not I think quote that famous passage from Lecky in which he shows how the prostitute, through centuries of suffering and ill-fame, has borne the curse and contempt of Society in order that her more fortunate sister might rejoice in the achievement of a pure marriage. The ideal of a monogamic union has been established in a sense directly by the slur cast upon the free woman. If, however, as many people think, a certain latitude in sexual relations is not only admissible but in the long run, and within bounds, desirable, it becomes clear that the prostitute is that person who against heavy odds, and at the cost of a real degradation to herself, has clung to a tradition which, in itself good, might otherwise have perished in the face of our devotion to the splendid ideal of the exclusive marriage. There has been a time in history when the prostitute (if the word can properly be used in this connection) has been glorified, consecrated to the temple-service and honored of men and gods (the hierodouloi of the Greeks, the kodeshoth and kodeshim of the Bible, &c.) There has also been a time when she has been scouted and reviled. In the future there will come a time when, as free companion, really free from the curse of modern commercialism, and sacred and respected once more, she will again be accepted by society and take her place with the rest.

And so with other cases. On looking back into history we find that almost every human impulse has at some age been held in esteem and allowed full play; thus man came to recognise its beauty and value. But then lest it should come (as it surely would) to tyrannise over the rest, it has been dethroned, and so in a later age the same quality is scouted and banned. Last of all it has to find its perfect human use and to take its place with the rest. Up to the age of Civilisation (according to writers on primitive Society) the early tribes of mankind, though limited each in their habits, were essentially democratical in structure. In fact nothing had occurred to make them otherwise. Each member stood on a footing of equality with the rest; individual men had not in their hands an arbitrary power over others; and the tribal life and standard ruled supreme. And when, in the future and on a much higher plane, the true Democracy comes, this equality which has so long been in abeyance will be restored, not only among men but also, in a sense, among all the passions and qualities of manhood: none will be allowed to tyrannise over others, but all will have to be subject to the supreme life of humanity. The chariot of

Man instead of two horses will have a thousand; but they will all be under control of the charioteer. Meanwhile it may not be extravagant to suppose that all through the Civilisation-period the so-called criminals are keeping open the possibility of a return to this state of society. They are preserving, in a rough and unattractive husk it may be, the precious seed of a life which is to come in the future; and are as necessary and integral a part of society in the long run as the most respected and most honored of its members at present.

The upshot then of it all is that "morals" as a code of action have to be discarded. There exists no such code, at any rate for permanent use. One age, one race, one class, one family, may have a code which the users of it consider valid, but only they consider it valid, and they only for a time. The Decalogue may have been a rough and useful ready-reckoner for the Israelites; but to us it admits of so many exceptions and interpretations that it is practically worthless. "Thou shalt not steal." Exactly; but who is to decide, as we saw at the outset, in what "stealing" consists? The question is too complicated to admit of an answer. And when we *have* caught our half-starved tramp "snaking" a loaf, and are ready to condemn him, lo! Lycurgus pats him on the back, and the modern philosopher tells him that he is keeping open the path to a regenerate society! If the tramp had also been a philosopher he would perhaps have done the same act not merely for his own benefit but for that of society, he would have committed a crime in order to save mankind.

There is nothing left but Humanity. Since there is no ever-valid code of morals we must sadly confess that there is no means of proving ourselves right and our neighbors wrong. In fact the very act of thinking whether *we* are right (which implies a sundering of ourselves, even in thought, from others) itself introduces the element of wrongness; and if we are ever to *be* "right" at all, it must be at some moment when we fail to notice it—when we have forgotten our apartness from others and have entered into the great region of human equality. Equality—in that region all human defects are redeemed; they all find their place. To love your neighbor *as* yourself is the whole law and the prophets; to feel that you are "equal" with others, that their lives are as your life, that your life is as theirs—even in what trifling degree we may experience such things—is to enter into another life which includes both sides; it is to pass beyond the sphere of moral distinctions, and to trouble oneself no more with them. Between lovers there are no duties and no rights; and in the life of humanity, there is only an instinctive mutual service expressing itself in whatever way may be best at the time. Nothing is forbidden, there is nothing which may not serve. The law of Equality is perfectly flexible, is adaptable to all times and places, finds a place for all the elements of character, justifies and redeems them all without exception; and to live by it is perfect freedom. Yet not a law: but rather as said, a new life, transcending the individual life, working through it from within, lifting the self into another sphere, beyond corruption, far over the world of Sorrow.

The effort to make a distinction between acting for self and acting for one's neighbor is the basis of "morals." As long as a man feels an ultimate antagonism between himself and society, as long as he tries to hold his own life as a thing apart from that of others, so long must the question arise whether he will act for self *or* for those others. Hence flow a long array of terms—distinctions of right and wrong, duty, selfishness, self-renunciation, altruism, etc. But when he discovers that there is no ultimate antagonism between himself and society; when he finds that the gratification of every desire which he has or can have may be rendered social, or beneficial to his fellows, by being used at the right time and place, and on the other hand that every demand made upon him by society will and must gratify some portion of his nature, some desire of his heart—why, all the distinctions collapse again; they do not hold water any more. A larger life descends upon him, which includes both sides, and prompts actions in accordance with an unwritten and unimagined law. Such actions will sometimes be accounted "selfish" by the world; sometimes they will be be accounted "unselfish"; but they are neither, or—if you like—both; and he who does them concerns himself not with the names that may be given to them. The law of Equality includes all the moral codes, and is the stand-point which they cannot reach, but which they all aim at.

Judged by this final standard then, it may doubtless fairly be said—since we all fall short of it—that we are all criminals, and deserve a good hiding; and even that some of us are greater criminals than others. Only of this real criminality the actual moral and legal codes afford but ineffectual tests. I may be a far worse or more self-included ("idiotic" or brutal) man than you, but the mere fact that I have violated the laws and been clapped into prison does not prove it. There may be, probably is, a real and eternal difference represented by the words Right and Wrong, but no statement that we can make will ever quite avail to define it. One use, however, of all these laws and codes in the past, imperfect though they were, may have been to gradually excite the consciousness in the individual of his opposition to society, and so prepare the way for a true reconcilement. As Paul says "I had not known sin, but by the law," and if we had not been cudgeled and bruised for centuries by this rough bludgeon of social convention we should not now be so sensitive as we are to the effect of our actions upon our neighbors, nor so ready for a social life in the future which shall be superior to law.

Of course the ultimate reconcilement of the individual with society—of the unit Man with the mass-Man—involves the subordination of the desires, their subjection to the true self. And this is a most important point. It is no easy lapse that is here suggested, from morality into a mere jungle of human passion; but a toilsome and long ascent—involving for a time at any rate a determined self-control—into ascendancy over the passions; it involves the complete mastery, one by one, of them all; and the recognition and allowance of them only because they are mastered. And it is just this training and subjection of the passions—as of winged horses which are to draw the human chariot—which necessarily forms such a long and painful process of human evolution. The old moral codes are a part of this process; but they go on the plan of extinguishing some of the passions—seeing that it is sometimes easier to shoot a restive horse than to ride him. We however do not want to be lords of dead carrion but of living powers; and every steed that we can add to our chariot makes our progress through creation so much the more splendid, providing Phoebus indeed hold the reins, and not the incapable Phaeton.

And by becoming thus one with the social self, the individual instead of being crushed is made far vaster, far grander than before. The renunciation (if it must be so called) which he has to accept in abandoning merely individual ends is immediately compensated by the far more vivid life he now enters into. For every force of his nature can now be utilised. Planting himself out by contrast he stands all the firmer because he has a left foot as well as a right, and when he acts, he acts not half-heartedly as one afraid, but, as it were, with the whole weight of Humanity behind him. In abandoning his exclusive individuality he becomes for the first time a real and living individual; and in accepting as his own the life of others he becomes aware of a life in himself that has no limit and no end. That the self of any one man is capable of an infinite gradation from the most petty and exclusive existence to the most magnificent and inclusive seems almost a truism. The one extreme is disease and death, the other is life everlasting. When the tongue for example—which is a member of the body—regards itself as a purely separate existence for itself alone, it makes a mistake, it suffers an illusion, and descends into its pettiest life. What is the consequence? Thinking that it exists apart from the other members, it selects food just such as shall gratify its most local self, it endeavors just to titillate its own sense of taste; and living and acting thus, ere long it ruins that very sense of taste, poisons the system with improper food, and brings about disease and death. Yet if healthy how does the tongue act? Why, it does not run counter to its own sense of taste, or stultify itself. It does not talk about sacrificing its own inclinations for the good of the body and the other members; but it just acts as being one in interest with them and they with it. For the tongue *is* a muscle, and therefore what feeds it feeds all the other muscles; and the membrane of the tongue *is* a prolongation of the membrane of the stomach, and that is how the tongue knows what the stomach will like; and the tongue *is* nerves and blood, and so the tongue may act for nerves and blood all over the body, and so on. Therefore the tongue may enter into a wider life than that represented by the mere local sense of taste, and experiences more pleasure often in the drinking of a glass of water which the whole body wants, than in the daintiest sweetmeat which is for itself alone.

Exactly so man in a healthy state does not act for himself alone, practically cannot do so. Nor does he talk cant about "serving his neighbors," &c. But he simply acts for them as well as for himself, because they are part and parcel of his life—bone of his bone and flesh of his flesh; and in doing so he enters into a wider life, finds a more perfect pleasure, and becomes more really a man than ever before. Every man contains in himself the elements of all the rest of humanity. They lie in the back-ground; but they are there. In the front he has his own special faculty developed—his individual façade, with its projects, plans and purposes: but behind sleeps the Demos-life with far vaster projects and purposes. Some time or other to every man must come the consciousness of this vaster life.

The true Democracy, wherein this larger life will rule society from within—obviating the need of an external government—and in which all characters and qualities will be recognised and have their freedom, waits (a hidden but necessary result of evolution) in the constitution of human nature itself. In the pre-Civilisation period these vexed questions of "morals" practically did not exist; simply because in that period the individual was one with his tribe and moved (unconsciously) by the larger life of his tribe. And in the post-Civilisation period, when the true Democracy is realised, they will not exist, because then the man will know himself a part of humanity at large, and will be consciously moved by forces belonging to these vaster regions of his being. The moral codes and questionings belong to Civilisation, they are part of the forward effort, the struggle, the suffering, and the temporary alienation from true life, which that term implies.

EXFOLIATION.

"Creation's incessant unrest, exfoliation."

WHITMAN.

I THINK it may perhaps be agreed, once for all, that the human mind is incapable of really defining even the smallest fact of nature. The simplest thing, or event, baffles us at the last. It is like trying to look at the front and back of a mirror at the same time. The utmost squinting avails not. The ego and the non-ego dance eluding through creation. To catch them both in any mortal object and pin them there, surpasses our powers. And yet they are there. Montaigne quotes somewhere the words of S. Augustine: *modus, quo corporibus adhaerent spiritus . . . omnino mirus est, nec comprehendi ab homine potest; et hoc ipse homo est.* "The manner whereby spirits adhere to bodies is altogether wonderful, and cannot be conceived of by men; and yet this is man." Man himself contains, or rather *is*, the reconcilement of this and numberless other contradictions. We actually every day perform and exhibit miracles which the mental part of us is utterly powerless to grapple with. Yet the solution, the intelligent solution and understanding of them *is* in us; only it involves a higher order of consciousness than we usually deal with—a consciousness possibly which includes and transcends the ego and the non-ego, and so can envisage both at the same time and equally

—a fourth-dimensional consciousness to whose gaze the interiors of solid bodies are exposed like mere surfaces—a consciousness to whose perception some usual antitheses like cause and effect, matter and spirit, past and future, simply do not exist. I say these higher orders of consciousness are in us waiting for their evolution; and, until they evolve, we are powerless really to understand anything of the world around us.

Meanwhile, since we *must* have formulæ and generalisations to think by, we are fain to accept our local views, and look on the world from this side or from that. Sometimes we are idealists, sometimes we are materialists; sometimes we believe in mechanics, sometimes in human or spiritual forces. The science of the last fifty years has, as pointed out in a preceding paper, looked at things more from the mechanical than the distinctively human side—from the point of view of the non-ego, rather than of the ego. Reacting from an extreme tendency towards a subjective view of phenomena, which characterised the older speculations, and fearing to be swayed by a kind of partiality towards himself, the modern scientist has endeavored to remove the human and conscious element from his observations of Nature. And he has done valuable work in this way—but of course has been betrayed into a corresponding narrowness.

In fact the main scientific doctrine of the day, Evolution, is obviously suffering from this treatment, and the following remarks are merely a few notes by way of suggestion of some things which may be said on its more specially human side. For since each man is a part of nature, and in that sense a part also of the evolution-process, his own subjective experience ought at least to throw some light on the conditions under which evolution takes place, and to contribute something towards an understanding of the problem.

If the question is: What is the cause of Variation among

animals? some approximation towards an answer ought to be got by each person asking himself, "Why do I vary?" Why—he might say—am I a different person from what I was ten years ago, or when I was a boy? Why have I varied in one direction and my brothers and sisters from the same nest in other directions? Though my individual consciousness only covers the small ground of my own life, and does not extend back to that of my father or forward to that of my son, still the intimate knowledge that I have of the forces acting on me during that short period may help me to an understanding of the forces that bring about the modification of men and animals at large, and the discovery of some laws of my own growth may reveal to me the laws also of race-growth.

In answer to such a question, it would speedily appear that there were two general causes determining direction of change or growth in the individual, which might be conveniently distinguished from each other—an external and an internal. In the first place the supposed person might say, "External conditions forced me along these lines. My father was a town artisan, but he apprenticed me to a farmer. I grew up a farmer's boy, and became an agricultural type as you see. I did not particularly care for farming, sometimes indeed I would have been glad to be out of it; but practically I succumbed to circumstances, and here I am." But in the second place he might answer thus:—"My father was himself a farmer; I was early used to the craft, and should no doubt have grown up in it, had I not hated it like poison. I loved music, broke away from home, joined a band, got on the musical staff of a small theatre, and am now a professional musician. My frame is comparatively slight, and my hands are of the nervous type, as you see. Of course I have some of the old agricultural stock left in me, but I feel that that is dying out." The one cause would be a change of external conditions, forcing the man to accommodate himself to them; the other would be a change of internal conditions, an inward growth, expressing itself first in the form of an intense desire, and compelling the man to change himself and probably also his environment in obedience to it. Two such general sets of causes, I say, could be roughly distinguished from each other; and probably indeed are recognised less or more distinctly by everyone as acting to modify his life. Nor can the life of a man at any time be said to be ruled by one of these forces alone. No man is modified by external conditions alone, without any play or reaction of inner needs and desires and growth from within; nor is any man transformed in obedience to an inner expansion without sundry lets and hindrances from without. The two forces are in constant play upon one another; but in some ways that would appear to be the more important which proceeds from the Man (or creature) himself, since this is obviously vital and organic to him, and therefore the most consistent and reliable factor in his modification, while the external force—arising from various and remote causes—must rather be regarded as discontinuous and accidental.

I propose, therefore, in these few pages to consider especially this inner force producing modification in man and animals—to try and find out of what nature it is, what is the law, and what are the limits of its action—premising always, as already suggested, that this distinction between "inner" and "outer," which is convenient and easy to handle on certain planes of thought, may ultimately, and in the last resort, prove very difficult or even impossible to maintain.

It is often said by Biologists that *function precedes organisation*—that is, man fights with his fellows before he makes weapons to fight with; the rudimentary animal digests food (as in the case of the amoeba) before it acquires a stomach or organ of digestion; it sees or is sensitive to light before it grows an eye; in society letters are carried by private hands before an organised postal system is created. Such facts properly considered are of vital importance. They show us, as it were by a sign-post, the direction of creation. They show how any new thing or modification of an old thing may come into being. They may be supplemented by a second statement—namely that *desire precedes function*. That is, man desires to injure his fellow before he actually fights with him; he experiences the wish to communicate with distant friends before ever he thinks of sending such a thing as a letter; the amoeba craves for food first, and circumvents its prey afterwards. Desire, or inward change, comes first, action follows, and organisation or outward structure is the result.

In man this "order of creation," if it may so be called, *i.e.*, from within outwards, is very marked. Whenever a man creates anything new he pursues it; when he builds a house for instance, or composes a poem or piece of music, or designs an Alpine tunnel, or whatever it may be. The order seems to be: first, a feeling—a dim want or desire; then the feeling becomes conscious of itself, takes shape in thought; the thought becomes more defined and issues in a distinct plan; the plan is committed to paper, models are made, &c.; and finally the actual work is begun and completed. The process appears as a movement from within outwards—the earliest and most authentic discernible source of the movement being a feeling—(though there may lie something behind that). Even in ordinary action the same order is manifest; for though of course *every* action is not preceded by desire—since we know that actions soon become habitual and more or less unconscious—still a vast number of them are immediately so preceded; and in the case of any action that is *new*, either to the individual or to the race, its inception is generally accompanied by effort so painful that it would not be exerted unless the desire were very strong. The difficulty which a man experiences in learning any new art, and the records of the many failures, struggles, oppositions, persecutions, &c., which have attended every new invention or innovation of any kind in human history afford plenty of evidence of this last point. Certainly the effort that accompanies a new action is not always faced so much from sheer desire of the new thing itself as from fear perhaps of something else—as it may be contended that monkeys did not take to climbing trees because they loved trees, but because they feared the beasts below, or that the giraffe did not stretch its neck because it particularly desired to feed on leaves, as because it could not get food any other way—but still, even in these cases the desire may be said to exist, though it is secondary—being founded upon another and more elementary desire—the desire namely of escaping pain or obtaining food. In either case a desire of some kind is a precedent condition of the new action. And so as we know of no case of a new action coming into play without being preceded by desire, we seem to be justified in supposing that all our actions when they were first initiated (in our forefathers if not in ourselves) were so preceded. If this is so, then since function is always preceded by desire, and organisation is preceded by function, organisation must necessarily be preceded by desire. And if this is the order of creation in man, should we not reasonably look in this direction for the key to the variation of animals and the order of creation in general? [1]

If a farmer's son is occasionally born who hates farming and loves music, and who ultimately through the force of his

[1] This does not of course preclude the action of external conditions, or imply that organisation is determined by desire *alone*. In fact organisation may be regarded as the expression of desire acting under conditions—as in the cases of the monkey and giraffe above.

desire (driving him into oppositions and difficulties and penurious struggles) transforms himself into a musician, is it not also likely that occasionally an animal is born who hates the customs of his tribe, and at last (also through struggles) transforms himself into something else? Even if he does not succeed (the animal) in entirely transforming himself, he likely transmits the desire in some degree to his descendants, and the transformation is thus carried on and completed later. For everywhere among the animals there *is* desire, of some kind or another, obviously acting; and if in man, by our own experience, desire is the precursor and first expression of growth, is there any reason why it should not also be so among animals? Lamarck gives the instance—among others—of a gasteropod; how the need or desire of touching bodies in front of it as it crawled along would result in the formation of tentacles. The gasteropod, he says, would keep making efforts to feel with the front of its head, and the determination of consciousness that way would be accompanied by a supply of nervous and other fluids, which would nourish the part and cause growth there—the *form* of the growth continuing in the same way to be determined by need—till at last two or more tentacles would appear. True, the inward determinations of consciousness may not be so vivid and varied in animals as they are in men; but they are persistent, and by the very cumulative force of habit which is so strong in animals, must at length penetrate down through function into organisation and external form. Who shall say that the lark, by the mere love of soaring and singing in the face of the sun, has not altered the shape of its wings, or that the forms of the shark or of the gazelle are not the long-stored results of character leaning always in certain directions, as much as the forms of the miser or the libertine are among men?

Such modification as this is very different from the "survival of the fittest" of the Darwinian evolution-theory. We may fairly suppose that both kinds of modification take place; but the latter is a sort of easy success won by an external accident of birth—a success of the kind that would readily be lost again; while the former is the uphill fight of a nature that has grown inwardly and wins expression for itself in spite of external obstacles—an expression which therefore is likely to be permanent. If the progenitors of man took to going upright on two legs instead of on all fours, merely because a few of them by *chance* were born with a talent for that position, which enabled them to escape the fanged and pursuing beasts, then when this danger was removed they might have plumped down again into the old attitude; but if the change was part and parcel of a true evolution—a true *unfolding* of a higher form latent within—an organic growth of the creature itself, then, though the moment of the evolution of this particular faculty might be determined by the fanged beasts, the fact of such evolution could not be determined by them. Besides are we to suppose that Man, the lord and ruler of the animals, came merely by way of *escape* from the animals? Do lords and rulers generally come so? Was it fear that made him a man? Were it not likelier that in that case he would have turned into a worm? He would have escaped better perhaps that way. Is it not rather probable that it was some nobler power that worked transforming—some dim desire and prevision of a more perfect form, the desire itself being the first consciousness of the urge of growth in that direction—that prompted him to push in the one direction rather than the other when he had to hold his own against the tigers? In fact is it not thus to-day, when a man has to meet danger, that the ideal which he has within him determines *how* he shall meet that danger, and others like it, and so ultimately determines the whole attitude and carriage of his body?

On the whole then, judging from man himself (and it seems most cautious and scientific to derive our main evidence from the being that we are best acquainted with), it certainly seems to me that though the external conditions are a very important factor in Variation, the central explanation of this phenomenon should be sought in an inner law of Growth—a law of expansion more or less common to all animate nature. Partly because, as said before, the unfolding of the creature from its own needs and inward nature is an organic process, and likely to be persistent, while its modification by external causes must be more or less fortuitous and accidental and sometimes in one direction and sometimes in another; partly also because the movement from within outwards seems to be most like the law of creation in general. Under this view the external conditions would be considered a secondary—though important cause of modification; and regarded rather as the influences that give form and detail to the great primal impulse of growth from within; while the creature's own ingenuity and good luck would occupy the ground between the two—as the means whereby the external conditions in each individual case would be turned to account to satisfy the inner needs, or the inner life would be accommodated to the external conditions.

If we take the external view of Variation—which is the one most favored by modern science—modification or race-growth appears as an unconscious or accretive process—similar to the formation of a coral reef. There is no line of growth native in the race itself, but at any moment it is supposed to have an equal tendency to vary in any direction. Surrounding conditions act selectively; and by a process of weeding out certain types survive; small successive modifications are thus accumulated; and gradually and in the lapse of ages a more pliable and differentiated creature, and more adaptable to a variety of conditions, is produced—in whom however mind is incidental, and has played but small part in the creature's evolution. This in the main is the Darwinian-evolution theory.

If we take the internal view, growth is from the first eminently conscious. Every change begins in the mental region—is felt first as a desire gradually taking form into thought, passes down into the bodily region, expresses itself in action (more or less dependent on conditions), and finally solidifies itself in organisation and structure. The process is not accretive but exfoliatory—a continual movement from within outwards. When the desire or mental condition which at first was painfully conscious, has overcome opposition and established itself in altered bodily structure, it has done its work, and becomes unconscious—the bodily function continuing for a long period to act automatically, till finally it is thrown off to make room for some later development. Thus race-growth or Variation is a process by which change begins in the mental region, passes into the bodily region where it becomes organised, and finally is thrown off like a husk. This may be called the theory of Exfoliation.

To illustrate our meaning. Let us take the development of an eye. In the amoeba there is a dim pervasive sensitiveness to light over the whole body, but there is no eye, nothing that we should call vision. Still this vague sensitiveness is of use to the amoeba. The shadow of its prey falling upon the creature and exciting a sensation hardly yet differentiated from touch helps to guide its movements. On this dim sensation it relies to some extent; its attention is directed towards it. Gradually, and in some descendant form, there comes to be a point on the body on which this attention is most specially concentrated. The faculty is localised; and from that moment a change is effected there, a differentiation and a special structure; everything that favors sensitiveness is encouraged at that place, everything that dulls it is removed; and before long—there is a rudimentary eye. To-day we use our perfected eyes, and are hardly conscious that we are doing so; but every power of vision that we have was thus won for us by some lowlier creature, step by step, with effort and with concentration. Or to take an illustration from society. To-day society is ill at ease; a dim feeling of discontent pervades all ranks and classes. A new sense of justice, of fraternity, has descended among us, which is not satisfied with mere chatter of demand and supply. For a long time this new sentiment or desire remains vague and unformed, but at last it resolves itself into shape; it takes intellectual form, books are written, plans formed; then after a time definite new organisations, for the distinct purpose of expressing these ideas, begin to exist in the body of the old society; and before so very long the whole outer structure of society will have been reorganised by them. After a few centuries the ideas for whose realisation we now fight and struggle with an intense consciousness will have become commonplace accepted institutions, more or less effete and ready to succumb before fresh mental births taking place from within.

The modern evolution theory would maintain that among many amoebas and descendant forms, one would at last by chance be born having the usual sensitiveness localised in a particular spot, and, surviving by force of this advantage, would transmit this "eye" to its posterity; or that in the progress of society, new economic conditions having arisen, that people would prosper best which most effectually and rapidly adapted itself to them. But though there is doubtless truth in this view, yet it seems when all has been said to be inadequate and even feeble; it omits at least one half of the problem. If we look at ourselves, as already pointed out, we see the two forces—the inner and the outer—acting and reacting on each other. May it not be so in animals? Lamarck, poorly off, blind, derided, was a true poet. "Animals vary from low and primitive types chiefly by dint of wishing"—and the world laughed and still laughs. But it was his deep sympathy even with the worms and insects (which he studied till he could discern them with his mortal eyes no longer) that led Lamarck to see the human nature and the human laws that moved within them; and as his outward sight grew dim there arose before him the inward vision of the true relationship which binds together all living creatures—which was indeed a vision of divine things, and as different from the mere mechanism-theory of the survival of the fittest as the sight of the starry heavens is different from a governess's lesson on the use of the globes.

On the theory of Exfoliation, which was practically Lamarck's theory, there is a force at work throughout creation, ever urging each type onward into new and newer forms. This force appears first in consciousness in the form of *desire.* Within each shape of life sleep wants without number, from the lowest and simplest to the most complex and ideal. As each new desire or ideal is evolved, it brings the creature into conflict with its surroundings, then gaining its satisfaction externalises itself in the structure of the creature, and leaves the way open for the birth of a new ideal. If then we would find a key to the understanding of the expansion and growth of all animate creation, such a key may exist in the nature of desire itself and the comprehension of its real meaning. It is not certain that it can be found here; but it may be.

What then is desire in Man? Here we come back again, as suggested at the outset, to Man himself. Though we see pretty clearly that desire is at work in the animals, and that it is the same in kind as exists in man, still among the animals it is but dim and inchoate while in man it is developed and luminous; in ourselves too we know it immediately, while in the animals only by inference. For both reasons therefore if we want to know the nature of desire—even to know its nature among animals—we should study it in Man. What then is desire—what is its culmination and completion—in Man? Practically it is love. Love is the sum and the solution of all desires in Man—that in which they converge; the interpretation of them; for which they all exist, and without which they would be considered useless. The more you look into this matter, the plainer it becomes. The other desires—the self-preservation desires—hunger, thirst, the desire of power—exist, but when they are satisfied they empty themselves into this one; they find their interpretation in it. The other desires are nothing by themselves—the most absorbing, avarice, ambition, desire of knowledge, taken alone, stultify themselves—but love perpetuates itself: it is a flame which uses all the rest as its fuel. And what is Love? It appears to us as a worship of and desire for the human form. In our bodies it is a desire for the bodily human form; in our interior selves it is a perception and worship of an ideal human form, it is the revelation of a Splendor dwelling in others, which—clouded and dimmed as it inevitably may come to be—remains after all one of the most real, perhaps the most real, of the facts of existence. Desire, therefore—as it exists in man—look at it how you will—as it unfolds and its ultimate aim becomes clearer and clearer to itself, is seen to be the desire and longing for the perfect human Form. May it not, must it not, be the same thing in animals and all thro' creation? Beginning in the most elementary and dim shapes, does it not grow through all the stages of organic life clearer and more and more powerful, till at last it attains to self-consciousness in humanity and becomes avowedly the leading factor in our development.

The desire which runs through creation is one desire. Rudimentary at first and hardly conscious of itself, throwing out a tentacle here, a foot there, developing an eye, a claw, a nostril, a wing, it seeks in innumerable shapes and with ever partial success to realise the image it has dimly conceived. The animal kingdom is the gymnasium, the school, the ante-chamber, of humanity; to walk thro' a zoological garden is to see the inchoate types of man, perched on branches, or browsing grass, or boring holes in the ground; it is to witness a grand rehearsal of some stupendous part, whose character we do not even yet fully see or understand. From such half-conscious beginnings the desire grows, its aim becomes clearer, till in the higher animals—the horse, the dog, the elephant, the bird, and many others—it becomes a marked and unmistakeable force drawing them close to man, uniting them to him in a kind of acknowledged kinship, and as obviously at work modifying their structure as can be. Finally in man himself it becomes an absorbing power; love becomes a conscious worship of the divine form; generation itself is the means whereby, in time, the supreme object of desire is realised. When at last the perfect Man appears, the key to all nature is found, every creature falls into its place and finds its Interpreter, and the purpose of creation is at last made manifest.

The Theory of Exfoliation then differs from that very specialised form of Evolution which has been adopted by modern science, in this particular among others: that it fixes the attention on that which appears last in order of Time, as the most important in order of causation, rather than on that which appears first; and recalls to us the fact that often in any succession of phenomena, that which is first in order of precedence and importance is the last to be externalised. Thus in the growth of a plant we find leaf after leaf appearing, petal within petal—a continual exfoliation of husks, sepals, petals, stamens and what-not; but the object of all this movement, and that which in a sense sets it all in motion, namely the seed, is the very last thing of all to be manifested. Or when a volcano breaks out—first of all we have a cracking and upheaval of superficial layers of ground, then of layers below these, then the outflow of lava, and *last of all* the uprush of the inner fires and forces which set it all agoing. What appears first in time, or in the outer world is—in the case of the building of a house, the making of bricks; in the case of the flower, the outermost bracts; in the case of a volcano, the stirring of the surface of the ground; and in the case of Life on the Earth, the appearance of protoplasms and primordial cells. The bricks are not the cause of the house (if indeed the word "cause" should be used here at all) but rather the house—or the conception of the house—is the cause of the bricks; and the cells are not the origin of Man, but Man is the original of the cells. The rationale of sea-anemones and mud-fish and flying foxes and elephants has to be looked for in man; he alone underlies them. And man is not a vertebrate because his ancestors were vertebrate; but the animals are vertebrate, because or in so far as they are fore-runners and offshoots of Man.

It has been frequently said that great material changes are succeeded by intellectual and finally by moral revolutions—as the conquests of Alexander passed on into the literary expansion of the Alexandrian schools and thence into the establishment of Christianity, or as the mechanical developments of our own time have been followed by immense literary and scientific activities, and are obviously passing over now into a great social regeneration; but a reconsideration of the matter might, I take it, lead us not so much to look on the later changes as *caused* by the earlier, as to look on the earlier as the indications and first outward and visible signs of the coming of the later. When a man feels in himself the upheaval of a new moral fact he sees plainly enough that that fact cannot come into the actual world all at once—not without first a destruction of the existing order of society—such a destruction as makes him feel satanic; then an intellectual revolution; and lastly only, a new order embodying the new impulse. When this new impulse has thoroughly materialised itself, then after a time will come another inward birth, and similar changes will be passed through again. So it might be said that the work of each age is not to build *on* the past, but to rise *out* of the past and throw it off; only of course in such matters where all forms of thought are inadequate it is hard to say that one way of looking at the subject is truer than another. As before, we should endeavor to look at the thing from different sides.

We are obliged to use images to think by—*e.g.* the opening of a flower or the accretive growth of a coral reef—and possibly it would save a good deal of trouble if we did not disguise by long words the truth that all our theories in science and philosophy are simply metaphors of this kind—but the *fact* still lies behind and below them.

Perhaps if we are to use the word Cause at all we should do well to use it in the old sense in which the *final* cause and the *efficient* cause are one (the *eidos* of Aristotle)—to use it not so much to link phenomena or externals to *each other* as to link each phenomenon in a group to the thought or feeling which underlies that group. The notes in the Dead March in Saul, for instance. We cannot say that one note is the cause of another, but we might say that each note stands in a causal subordination to the feeling which inspired the piece—which is the *origin* of the piece and the *result* of its performance—the alpha and omega of it. Similarly the ground floor in a house is not the cause of the first floor, nor the first floor of the second floor, nor that of the roof; but these actualities and the whole house itself stand in strict relationship to a mental something which is not in the same plane with them at all, nor an actuality in the same sense.

According to this view the notion that one configuration of atoms or bodies determines the next configuration turns out to be illusive. Both configurations are determined by a third something which does not belong to quite the same order of existence as the said atoms or bodies. Chance "laws" of succession may doubtless be found among physical events, and are valuable for practical purposes, but at any moment—owing to their superficiality—they may fail. Thus an insect observing the expansion of the petals of a chrysanthemum might frame a law of their order of succession in size and color, which would be valid for a time, but would fail entirely when the stamens appeared. Or, to take another illustration, physical science acts like a man trying to find direct causal relations between the various leaves of a tree, without first finding the relations of these to the branches and trunk—and so solving the problem indirectly. It deals only with the *surface* of the world of Man.

In thinking about such matters, Music, as Schopenhauer shows, is wonderfully illustrative; because in creating music man recognises that he is creating a world of his own—apart from and not to be confused with that other world of Nature (in which he does not recognise any of his handiwork). Supposing a non-musical person were to examine and analyse the score of a Beethoven symphony, he would be in the same position as a man examining and analysing Nature by purely scientific or intellectual methods. He would discover the recurrence of certain groups among the notes, he would establish laws of their sequences, would make all kinds of curious generalisations about them, and point out some remarkable exceptions, would even very likely be able to predict a bar or two over the page; his treatise would be very learned, and from a certain point of view interesting also, but how far would he be from any real understanding of his subject? Let him change his method: let him train his ear, let him hear the symphony performed, over and over, till he understands its meaning and knows it by heart; and then he will know at any rate something of why each note is there, he will see its fitness and feel in himself the "law" of its occurrence, and possibly in some new case will be able to predict several bars over the page! The symphony is not understood by examination and comparison of the notes alone, but by *experience* of their relation to deepest feelings; and Nature is not explained by laws, but by its becoming—or rather being felt to be—the body of Man; marvellous interpreter and symbol of his inward being.

There is a kind of knowledge or consciousness in us—as of our bodily parts, or affections, or deep-seated mental beliefs—which forms the base of our more obvious and self-conscious thought. This systemic knowledge grows even while the brain sleeps. It is not by any means absolute or infallible, but it affords, at any moment in man's history, the axiomatic ground on which his thought-structures, scientific and other, are built. Thus the axioms of Euclid are part of our present systemic knowledge, and afford the ground of all our geometry structures. But as the systemic consciousness grows, the ground shifts and the structures reared upon it fall. All our modern science, for instance, is founded on the acceptation of mechanical cause and effect as a basic fact of consciousness; but when that base gives way the entire structure will cave in, and a new edifice will have to be reared. Similarly, when the human form becomes distinctly visible to us in the animals—as an unavoidable part of our consciousness—this consciousness will form a new base or axiom for all our thought on the subject, and the theory of evolution, as hitherto conceived by science, will be entirely transformed.

Thus, although the experimental investigatory coral-reef accretion method of modern science is very valuable within its range, it must not be forgotten that the human mind does not progress more than temporarily by this method—that its progression is a matter of growth from within, and involves a continual *breaking away of the bases* of all thought-structures; so that while this latter—*i.e.*, the progression of the systemic consciousness of man—is necessary and continuous, the rise and fall of his thought-systems is accidental, so to speak, and discontinuous.

It is then finally in Man—in our own deepest and most vital experience—that we have to look for the key and explanation of the changes that we see going on around us in external Nature, as we call it; and our understanding of the latter, and of History, must ever depend from point to point on the exfoliation of new facts in the individual consciousness. Round the ultimate disclosure of the ideal Man, all creation (hitherto groaning and travailing towards that perfect birth) ranges itself, as it were like some vast flower, in concentric cycles; rank beyond rank; first all social life and history, then the animal kingdom, then the vegetable and mineral worlds. And if the outer circles have been the first in fact to show themselves, it is by this last disclosure that light is ultimately thrown on the whole plan; and, as in the myth of the Eden-garden, with the appearance of the perfected human form that the work of creation definitely completes itself.

CUSTOM.

"Whatever is off the hinges of custom is believed to be also off the hinges of reason; though how unreasonably, for the most part, God knows."

MONTAIGNE.

EVERY human being grows up inside a sheath of custom, which enfolds it as the swathing clothes enfold the infant. The sacred customs of its early home, how fixed and immutable they appear to the child! It surely thinks that all the world in all times has proceeded on the same lines which bound its tiny life. It regards a breach of these rules (some of them at least) as a wild step in the dark, leading to unknown dangers.

Nevertheless its mental eyes have hardly opened ere it perceives, not without a shock, that whereas in the family dining-room the meat always precedes the pudding, below-stairs and in the cottage the pudding has a way of coming before the meat; that whereas its father puts the manure on the top of his seed-potatos in spring, his neighbor invariably places his potatos on top of the manure. All its confidence in the sanctity of its home life and the truth of things is upset. Surely there must be a right and a wrong way of eating one's dinner or of setting potatos, and surely, if any one, "father" or "mother" must know what is right. The elders have always said (and indeed it seems only reasonable) that by this time of day everything has been so thoroughly worked over that the best methods of ordering our life—food, dress, domestic practices, social habits, &c., have long ago been determined. If so, why these divergencies in the simplest and most obvious matters?

And then other things give way. The sacred seeming-universal customs in which we were bred turn out to be only the practices of a small and narrow class or caste; or they prove to be confined to a very limited locality, and must be left behind when we set out on our travels; or they belong to the tenets of a feeble religious sect; or they are just the products of one age in history and no other. And the question forces itself upon us, Are there really no natural boundaries? has not our life anywhere been founded on reason and necessity, but only on arbitrary habit? What is more important than food, yet in what human matter is there more unaccountable divergence of practice? The Highlander flourishes on oatmeal, which the Sheffield ironworker would rather starve than eat; the fat snail which the Roman country gentleman once so prized now crawls unmolested in the Gloucestershire peasant's garden; rabbits are taboo in Germany; frogs are unspeakable in England; sauer-kraut is detested in France; many races and gangs of people are quite certain they would die if deprived of meat, others think spirits of some kind a necessity, while to others again both these things are an abomination. Every country district has its local practices in food, and the peasants look with the greatest suspicion on any new dish, and can rarely be induced to adopt it. Though it has been abundantly proved that many of the British fungi are excellent eating, such is the force of custom that the mushroom alone is ever publicly recognised, while curiously enough it is said that in some other countries where the claims of other agarics are allowed the mushroom itself is not used! Finally, I feel myself (and the gentle reader probably feels the same) that I would rather die than subsist on *insects*, such is the deep-seated disgust we experience towards this class of food. Yet it is notorious that many races of respectable people adopt a diet of this sort, and only lately a book has been published giving details of the excellent provender of the kind that we habitually overlook—tasty morsels of caterpillars and beetles, and so forth! And indeed, when one comes to think of it, what can it be but prejudice which causes one to eat the periwinkle and reject the land-snail, or to prize the lively prawn and proscribe the cheerful grasshopper?

It is useless to say that these local and other divergencies are rooted in the necessities of the localities and times in which they occur. They are nothing of the kind. For the most part they are mere customs, perhaps grown originally out of some necessity, but now perpetuated from simple habit and inherent human laziness. This can perhaps best be illustrated by going below the human to the kingdom of the animals. If customs are strong among men they are far stronger among animals. The sheep lives on grass, the cat lives on mice and other animal food. And it is generally assumed that the respective diets are the most "natural" in each case, and those on which the animals in question will readiest thrive, and indeed that they could not well live on any other. But nothing of the kind. For cats can be bred up to live on oatmeal and milk with next to no meat; and a sheep has been known to get on very comfortably on a diet of port wine and mutton chops! Dogs, whose "natural" food in the wild state is of the animal kind, are undoubtedly much healthier (at any rate in the domestic state) when kept on farinaceous substances with little or no meat, and indeed they take so kindly to a vegetable diet that they sometimes become perfect nuisances in a garden—eating strawberries, gooseberries, peas, &c. freely off the beds when they have once learned the habit. Any one in fact who has kept many pets knows what an astonishing variety of food they may be made to adopt, though each animal in the wild state has the most intensely narrow prejudices on the subject, and will perish rather than overstep the customs of its tribe. Thus pheasants will eat fern-roots in winter when snow covers the ground, but the grouse "don't eat fern-roots," and die in consequence. A wolf of an inquiring turn of mind would probably find strawberries and peas as good food as a dog does, but it is practically certain that any ordinary member of the genus would perish in a garden full of the same if deprived of his customary bones.

All this seems to indicate what an immensely important part mere custom plays in the life of men and animals. The main part of the power which man acquires over the animals depends upon his establishing habits in them which once established they never think of violating: and the almost insuperable nature of this force in animals throws back light on the part it plays in human life.

Of course, I am not contending in the above remarks upon food that there is no physiological difference between a dog and a sheep in the matter of their digestive organs, and that the one is not by the nature of its body more fitted for one kind of food than the other; but rather that we should not neglect the importance of mere habit in such matters. Custom changed first; the change of physiological structure followed slowly after. What happened was probably something like this. Some time in the far back past a group of animals, driven perhaps by necessity, took to hunting in packs in the woods; it developed a modified physical structure in consequence, and special habits which in the course of time became deeply fixed in the race. Another group saved its life by taking to grazing. Grass is poor food; but it was the only chance this group had, and in time it got so accustomed to eating grass that it could not imagine any other form of diet, and at first would refuse even oysters when placed in its way! Another group saw an opening in trees; it developed a long neck and became the giraffe. But the fact that the giraffe lives on leaves, and the sheep on grass, and the wolf on animal matter, and that custom is in each so strong that at first the creature will refuse any other kind of diet, does not of itself prove that that diet is the best or most physiologically suitable for it. In other words, it is an assumption to suppose that "adaptation to environment" is the sole or even the main factor in the constitution of well-marked varieties or genera; for this is to neglect (among other things) the force of mere use or wont, which has about the same import in race-growth that momentum has in dynamics; and causes the race, once started in any direction, to maintain its line of movement—and often in despite of its environment—even for thousands of years.

Returning to man we see him enveloped in a myriad customs—local customs, class customs, race customs, family customs, religious customs; customs in food, customs in clothing, customs in furniture, form of habitation, industrial production, art, social and municipal and national life, &c.; and the question arises, Where is the grain of necessity which underlies it all? How much in each case is due to a real fitness in nature, and how much to mere otiose habit! The first thing that meets my eye in glancing out of the window is a tile on a neighboring roof. Why are tiles made S-shaped in some localities and flat in others? Surely the conditions of wind and rain are much the same in all places. Perhaps far back there was a reason, but now nothing remains but—custom. Why do we sit on chairs instead of the floor, as the Japanese do, or on cushions like the Turk? It is a custom, and perhaps it suits with our other customs. The more we look into our life and consider the immense variety of habit in every department of it—even under conditions to all appearances exactly similar—the more are we impressed by the absence of any very serious necessity in the forms we ourselves are accustomed to. Each race, each class, each section of the population, each unit even, vaunts its own habits of life as superior to the rest, as the only true and legitimate forms; and peoples and classes will go to war with each other in assertion of their own special beliefs and practices; but the question that rather presses upon the ingenuous and inquiring mind is, whether any of us have got hold of much true life at all?—whether we are not rather mere multitudinous varieties of caddis-worms shuffled up in the cast-off skins and clothes and débris of those who have gone before us, with very little vitality of our own perceptible within? How many times a day do we perform an action that is authentic and not a mere mechanical piece of repetition? Indeed, if our various actions and practices were authentic and flowing from the true necessity, perhaps we shouldn't quarrel with each other over them so often as we do.

And then to come to the subject of morals. These also are customs—divergent to the last degree among different races, at different times, or in different localities; customs for which it is often difficult to find any ground in reason or the "fitness of things." Thieving is supposed to be discountenanced among us, yet our present-day trade-morality sanctions it in a thousand different forms; and the respectable usurer (who can hardly be said to be other than a thief) takes a high place at the table of life. To hunt the earth for game has from time immemorial been considered the natural birthright and privilege of man, until the landlord class (whom wicked Socialists now denounce!) invented the crime of poaching and hanged men for it. As to marriage customs, in different times and among different peoples, they have been simply innumerable. And here the sense of inviolability in each case is most powerful. The severest penalties, the most stringent public opinion, biting deep down into the individual conscience, enforce the various codes of various times and places; yet they all contradict each other. Polygamy in one country, polyandry in the next; brother and sister marriage allowed at one time, marriage with your mother's cousin forbidden at another; prostitution sacred in the temples of antiquity, trampled under foot in the gutters of our great cities of to-day; monogamy respectable in one land, a mark of class-inferiority in another; celibacy scorned by some sections of people, accepted as the highest state by others; and so on.

What are we to conclude from all this? Is it possible, once we have fairly faced the immense variety of human life in *every* department of arts, manners, and morals—a variety, too, existing in a vast number of cases under conditions to all intents and purposes quite similar—is it possible ever again to suppose that the particular practices which *we* are accustomed to are very much better (or, indeed, very much worse) than the particular practices which others are accustomed to? We have been born, as I said at first, into a sheath of custom which enfolds us with our swaddling-clothes. When we begin to grow to manhood we see what sort of a thing it is which surrounds us. It is an old husk now. It does not bear looking into; it is rotten, it is inconsistent, it is thoroughly indefensible; yet very likely we have to accept it. The caddis-worm has grown to its tube and cannot leave it. A little spark of vitality amid a heap of dead matter, all it can do is to make its dwelling a little more convenient in shape for itself, or (like the coral insect) to prolong its growth in the most favorable direction for those that come after. The class, the caste, the locality, the age in which we were born has determined our form of life, and in that form very likely we must remain. But a change has come over our minds. The vauntings of earlier days we abandon. *We*, at any rate, are no better than anybody else, and at best, alas! are only half alive.

If these, then, are our conclusions, is it not with justice that children and early races keep so rigidly to the narrow path that custom has made for them? Have they not an instinctive feeling that to forsake custom would be to launch out on a trackless sea where life would cease to have any special purpose or direction, and morality would be utterly gulfed? Custom for them is the line of their growth; it is the coral-branch from the end of which the next insect builds; it is the hardening bark of the tree-twig which determines the direction of the growing shoot. It may be merely arbitrary, this custom, but that they do not know; its appearance of finality and necessity may be quite illusive; but the illusion is necessary for life, and the arbitrariness is just what makes one life different from another. *Till he grows to manhood*, the human being, *he cannot do without it.*

And when he grows to manhood, what then? Why he dies, and so becomes alive. The caddis-fly leaves his tube behind and soars into the upper air; the creature abandons its barnacle existence on the rock and swims at large in the sea. For it is just when we die to custom that, for the first time, we rise into the true life of humanity; it is just when we abandon all prejudice of our own superiority over others, and become convinced of our entire indefensibleness, that the world opens out with comrade faces in all directions; and when we perceive how entirely arbitrary is the setting of our own life, that the whole structure collapses on which our apartness from others rests, and we pass easily and at once into the great ocean of freedom and equality.

This is, as it were, a new departure for man, for which even to-day the old world, overlaid with myriad customs now brought into obvious and open conflict with each other, is evidently preparing. The period of human infancy is coming to an end. Now comes the time of manhood and true vitality.

Possibly this is a law of history, that when man has run through every variety of custom a time comes for him to be freed from it—that is, he uses it indifferently according to his requirements, and is no longer a slave to t; all human practices find their use, and none are forbidden. At this point, whenever reached, "morals" come to an end and humanity takes its place—that is to say, there is no longer any code of action, but the one object of all action is the deliverance of the human being and the establishment of equality between oneself and another, the entry into a new life, which new life when entered into is glad and perfect, because there is no more any effort or strain in it; but it is the recognition of oneself in others, eternally.

Far as custom has carried man from man, yet when at last in the ever-branching series the complete human being is produced, it knows at once its kinship with all the other forms. "I have passed my spirit in determination and compassion round the whole earth, and found only equals and lovers." More, it knows its kinship with the animals. It sees that it is only habit, an illusion of difference, that divides; and it perceives after all that it is the same human creature that flies in the air, and swims in the sea, or walks biped upon the land.

THE NEED OF A RATIONAL AND HUMANE SCIENCE.[1]

IN bringing before you this subject of a Rational and Humane Science you will perhaps forgive me if I dwell for a few moments on some points of personal history in relation to it. After reading mathematics for some four years at Cambridge, it happened to me for the next ten years or so to be engaged in the study of the physical sciences, and in lectures on these subjects. Naturally, during the earlier part of this period I accepted the current methods and conclusions without any question. But as time went on I became aware of a certain dissatisfaction; I felt that many of the laws of Science, enounced as universal truths, were of very limited application only, that many of the conclusions, so strongly insisted on, were of quite doubtful validity; and at last this increasing dissatisfaction culminated in a rather violent attack or criticism of Modern Science which I wrote and published about the year 1884.[2]

Now, looking back, at this interval of time, though I admit that my attack was somewhat hasty and crude in detail, I feel

[1] Being a reprint of an address given before the Humanitarian League.

[2] Afterwards reprinted in a modified form, as "Modern Science—a Criticism," in the first edition (1889) of the present book.

that in its main contention it was thoroughly justified, and I do not feel the least inclined to withdraw it.

What was that main contention? It was as follows. Modern Science is an attempt (and no doubt it would accept this definition of itself) to survey and classify the phenomena of the world in the pure dry light of the intellect, uncoloured by feeling; and so far is an effort to separate the intellectual in man from the merely perceptive, the emotional, the moral, and so forth. It was in this very fact that my criticism lay; for I contended that such a separation was in the long run quite impossible.

But before proceeding to defend this position, let me admit at once that this attempt of Modern Science to get rid of human feeling and to look at everything in the dry light of the intellect was in some respects a very grand one. When you consider what the Old-time Science was, with its fancies and prejudices, its dragons pasturing upon the sun and moon in eclipses, its immolations of hundreds of human beings to appease some god of pestilence or earthquake, its panics, its superstitions, and its incapability of regarding anything except from the point of view of that thing's influence on man's own comfort and his little hopes and fears, it was indeed a grand advance to try and see *facts*, uncoloured and for themselves alone. It was an effort of Man as it were to rise above himself, to which I accord the fullest credit and honour.

And yet, during the time spoken of, it kept growing on me: first, that the attempt was an impossible one; secondly, that the Science so-called was not a true Science; and thirdly, that in its pretence to an intellectual exactitude which it did not really possess, this Modern Science was leading to a narrow-mindedness and a dogmatism as bad as the old.

There is in fact (so I think) a fallacy in the attempt. But how shall I describe it? Our relations to the world may, quite roughly speaking, be divided into three groups—those that are sensuous and perceptional, those that are purely intellectual, and those that are of an emotional and moral order. Take any object of Nature—a bird, for instance. We may look upon the bird as an object of sense-perceptions—its form, its colour, its song, and so forth. Some people attain to extraordinary skill and quickness in this department, recognising in a moment the note or even the flight of a songster. Then again we may look upon the bird from the intellectual side—we may study it in relation to its surroundings—the form of its wings, the length of its leg, the character of its beak, and their adaptation to its habits, to its locality, to its food, and so forth. Thus we may get a whole series of purely intellectual results—relations of the bird to the world in which it lives. This is the special field of the present-day Science. But, again, we may regard the bird in its emotional and moral relations to *us*. One man at the sight of it may be affected with admiration of its beauty, with tenderness towards it, or sympathy; another may be stimulated to wonder whether he can kill it, or whether it is good to eat! Modern Science is indifferent to what this last set of relations may be; it does not concern itself much with the first; but it takes the middle term, the purely intellectual, and seeks to abstract that from the others, to study the bird, or whatever the object may be, in the one aspect only. But can that really be done?—The answer is, of course, No.

To show my general meaning, and why I consider the claim an impossible one, let us imagine a little cell—one of the myriads which constitute the human body—professing in the same sort of way to stand outside the body and explain the laws of the other cells and the body at large. It is obvious that the little cell, swept along in the currents of the body and swayed by its emotions, in close proximity and contact with some portions of the organism, and far remote from others, cannot possibly pretend to any such impartial judgment. It is obvious not only that it would not have all the clues of the problem at its command, but that its own needs and experiences would prejudice it frightfully in the interpretation of such clues as it had. Yet man is such a little cell in the body of Nature, or, if you like, in the body of the Society of which he forms a part.

There is however one way, it seems to me, in which a cell in the human body *might* come to an adequate understanding of the body; and that would be rather through experience than through direct reasoning. It is conceivable that there might be some cell in the body which through the nerves, etc., was in actual touch and sympathetic relationship with every other cell. Then it certainly would have the materials of the required solution. Every change in other parts of the body would register itself in this particular cell; and its little brain (if it had one), without exactly making any great effort, would reflect sympathetically the structure of the whole body —would become, in fact, a mirror of it. This will perhaps give you the key to my notion of what a true Science might be.

But before proceeding to that, I want to go a little more in detail into the fallacy of the absolute intellectual view of Science. I say, first, that a complete summary of any object or process in Nature is impossible; secondly, that such summary as we do make is, and must inevitably and necessarily be, coloured by the underlying *feeling* with which we approach that phase of Nature.

To take the first point. You say, Why is a complete summary not possible? A watch or other machine may be completely described and defined; why should not (with a little more knowledge) a fir-tree, or the human eye, or the solar system, be completely described and defined?

And this brings us to what may be called the Machine-view of Science. It is curious (and yet I think it will presently be seen that it is quite what might have been expected) that during this last century or so, in which Machinery has played such an important part in our daily and social life, mechanical ideas have come to colour all our conceptions of Science and the Universe. Modern Science holds it as a kind of ideal (even though finding it at times difficult to realise) to reduce everything to mechanical action, and to show each process of Nature intelligible in the same sense as a Machine is intelligible. Yet this conception, this ideal, involves a complete fallacy. For the moment you come to think of it, you see that *no* part of Nature really even resembles a machine.

What is a machine in the ordinary sense? It is an aggregation of parts put together to fulfil certain definite actions and no others. A sewing-machine fulfils the purpose of sewing, a watch fulfils that of keeping time, and they fulfil those purposes only. All their parts subserve those actions, and in that sense may be completely described—as far as just their mechanical action is concerned—the same by a thousand mechanicians. But I make bold to say that *no* object in Nature fulfils just one action, or series of actions, and no others. On the contrary, every object fulfils an endless series of actions.

Let us take the Human Eye. And I choose this as an instance most adverse to my position, for there is no doubt that the Human Eye is one of the most highly specialised objects in creation. Helmholtz, as you know, is said to have remarked concerning it that if an Optician had sent him an instrument so defective he should have returned it with his compliments. Helmholtz was a great man, and I will not do him the injustice to suppose that he did not know what he was saying. He knew that, regarded as a machine for focussing rays of light, the eye was decidedly defective; but then he knew well enough, doubtless, *why* it was defective—namely, because it is by no means merely such a machine, but a great deal more.

The Eye, in fact, not only fulfils the action of focussing rays of light—like an Opera Glass or a Telescope—but it might be compared to another instrument, a Photographic Camera, in respect of the fact that it forms a picture of the outer world which it throws on a sensitive plate at the back—the Retina. But then, again, it is unlike any of these "machines," in the fact that it was never made by any Optician, human or divine, for any one definite purpose. On the contrary, as we know, it has grown, it has evolved; it has come down to us over the centuries, and over thousands and thousands of centuries, from dim beginnings in the lowliest organisms who first conceived the faculty of Sight, continually modified, continually shapen by small increments in various directions, in accordance with the myriad needs of a myriad creatures, living, some of them in water, some of them in air, requiring some of them to see at close quarters, some at great distances, some by one kind of light, some by another, and so forth. So that to-day it not only contains a great range of inherited, yet latent, faculties, but it is actually, in its complex structure, an epitome and partial record of its own extraordinary history.

As an instance of this last point, let me remind you that Sight was originally a differentiation of Touch. The light, the shadows, falling on the sensitive general surface of a primitive organism provoke a tactile irritation. In the course of evolution this sense specialises itself at some point of the surface into what we call Sight. Now, to-day, when the little picture formed by the fore-part of the Human Eye falls upon the Retina at the back, it falls upon a screen formed by the myriad congregated finger-tips, so to speak, of the optic nerve —the rods and cones, so-called—which cover like a mosaic the whole ground of the Retina, and *feel* with their sensitive points the images of the objects in the outer world. And so Sight is still Touch—it is the power of feeling or touching at a distance—as one sometimes in fact becomes aware in looking at things.

But then again on and beyond all these things—beyond the focussing and photographing of rays, beyond the latent adaptations to the needs of innumerable creatures, and the epitomising of ages of evolution—the Human Eye has faculties even more far-reaching perhaps and wonderful. It is the marvellous organ of human Expression. By the dilatations and contractions of the iris, by the altering convexities of the lens and the eyeball, and in a hundred other ways, it manages somehow to convey intelligence of Command, Control, Power, of Pity, Love, Sympathy, and all those myriad emotions which flit through the human mind—an endless series—a perfect encyclopædia. It is difficult even to imagine the eye without this power of language. And what other functions it may have it is not necessary to inquire. Highly specialised though it is, it is already obvious enough that to call it a Machine for focussing rays of light is monstrously and ludicrously inadequate—even as it would be to call the Heart (the very centre of emotion and life, and the symbol of human love and courage) a common Pump.

Nature is an infinitude, and can at no point be circumscribed by the human intellect. Nor obviously is there any sense in taking one little portion of Nature and isolating it from the rest, and then describing it exhaustively *as if* it really were so isolated. A thousand mechanicians will agree, as I have said, in their description of a machine, because in fact they will agree to view the machine just in the one aspect of its particular action; but ask a thousand people to describe one and the same face—or, better still, get a thousand portrait-painters, skilled in their art, to paint portraits of the same face—and you know perfectly well that all the likenesses will be different. And why will they be different? Simply because every face, however rude, has infinite sides, infinite aspects, and each painter selects what he paints from his own point of view. And the same is true of every object and process in Nature.

Then if these things are true (you ask again) how is it that scientific men *do* arrive at definite conclusions, and do agree with each other so far as they do?

It is, and obviously must be, by the method of isolation; by the method of selecting certain aspects of the problems presented to them, and ignoring others. For since *all* the relations of any phenomenon of Nature cannot possibly be compassed, the only way *must* be to ignore some and concentrate attention on others; and when there is a kind of tacit agreement as to which aspects shall be passed over and which considered, there is naturally an agreement in the results. Thus by this method, waiving all other aspects of the problem, the Eye may be described and defined as an optical instrument, the Heart as a common Pump, and the Solar System as a neat illustration of certain mechanical laws discovered by Galileo and Newton.

On the subject of the Solar System and Astronomy I will dwell for a few moments, as here—in this great example of the perfection of Modern Science—we have again a case apparently most adverse to my contention. The generalisations by which Newton established the nature of the planetary orbits have been a wonder to succeeding generations; the positions of the planets can be foretold, eclipses can be calculated with amazing accuracy. Yet every tyro in Mathematics knows that the equations which give these results can only be solved by what is called "neglecting small quantities" — that is, the problems cannot be solved in their entirety, but by leaving out certain terms and elements, which do not appear important, a solution can be approached. And naturally it has been an important point to show that these small quantities *may* be safely neglected. In the case, for instance, of the orbits of the planets round the sun, and of the moon round the earth, it was for a long time taken as proved that the small variations in the shape and position of each elliptic orbit would never be accompanied by any permanent increase or diminution in its *size*—that is, that the *mean distances* of the planets from the sun, and of the moon from the earth, would always remain within certain limits. Of late years however Professor George Darwin, taking up one of these poor little neglected quantities in the theory of the moon, found that it indicated after all very vast and very permanent, though of course very slow, changes in her mean distance from the earth; so that now it appears probable that the Moon's true orbit, instead of being a limited ellipse, is a continually though gradually enlarging Spiral, which may some day carry the Moon to a great distance from the earth. If an eclipse were calculated for twenty years in advance on the Elliptic theory or the Spiral theory, it would probably—so slow would be the divergence—make no perceptible difference; but in a hundred centuries the two theories would lead to results utterly different.

Thus the certitude of Astronomy as a Science arises largely from the fact that our *times* are so brief compared with Celestial periods. The proper periods of Celestial changes are to be reckoned by thousands, perhaps millions, of years;

but we, ignoring *that* aspect of the problem, fix our observations on one little point of time, and are quite satisfied with the result!

As another illustration of my meaning, consider the Fixed Stars, so-called. These stars in their groups and clusters, which we know so well by sight, have remained apparently in the very same, or nearly the same, relative positions during all the 2,000 or 3,000 years that we have any record of the shapes of the Constellations. Yet now by minute telescopic and spectroscopic examination we know that they are moving, and have been moving all the time, in various differing directions with great velocities, amounting to miles per second. Nevertheless, so great are the spaces concerned, so great the times, that all this long period has not sufficed to bring them into any greatly changed attitude with regard to each other! What would you think of an intelligent foreigner who, coming to England to study the game of cricket, remained on the cricket field for a quarter of a minute—during which time the players would have hardly changed their positions—and having noted a few points, went away and wrote a volume on the laws of the game? And what are we to think of poor little Man who, having noted the stars for a few centuries, is so sure that he understands their movements, and that he is versed in all the "ordinances of heaven."

Thus it would appear that every Nature-problem is so enormously complex that it can only be got at by what may be called the Method of Ignorance. Let us take a practical Science problem like that of Vaccination. The question here, put in its simplest terms, seems to be, Whether Vaccination, with calf or human lymph, prevents or alleviates Smallpox; and if it does, whether it does so without engendering other evils at least as great. At first sight this may appear to you a very simple question, and easy to solve; but the moment you come to think about it, you see its extreme complexity. In the first place, it is obvious that in a question like this, individual cases afford no test. It is obvious that the fact that A. is vaccinated and has not taken small-pox proves nothing, for there is nothing to show that he would have taken it if he had not been vaccinated. And when you have got people vaccinated by the hundred and the thousand, you still are not certain; for these people may belong to a certain class, or a certain locality, or may have certain habits and conditions of life, which may account for their comparative immunity, and these causes must be eliminated before any definite conclusion can be reached. Thus it is not till the great mass of the population is vaccinated that we can expect reliable statistics. But the introduction of a practice of this kind on so great a scale necessarily takes a long period of years, and meanwhile changes are taking place in the habits of the people, Sanitation is being improved, customs of Diet are altering, possibly (as so often happens in the history of an epidemic) the disease, having run its course, is beginning spontaneously to decline. And thus another series of possible causes has to be discussed.

Then, supposing the question, notwithstanding all these difficulties, to be so far settled in favour of the present system —there still arises that whole other series of difficulties with regard to the possibility of the spread of *other* diseases by the practice, and with regard to the *extent* of such spread, before we can arrive at any finale. This series of questions is almost as complex as the other; and it includes that great element of uncertainty—the question what interval of time may elapse between inoculation with a disease and its actual appearance. For if in several cases children break out with erysipelas immediately after vaccination, of course there is a certain presumption that vaccination has been the cause; but if the erysipelas only appears some years after, its connection with the operation may, though real, be impossible to trace.

The matter standing thus, it seems to us almost a mystery how it was that the medical authorities of the early days of Jennerism were so cocksure of their conclusions—until we remember that in arriving at those conclusions they practically *ignored* all these other points that I have mentioned, like changes of Sanitation, spontaneous decline of Small-pox, the spread of other diseases, etc., and simply limited themselves to one small aspect of the problem. But now, after this interval of time, when the neglected facts and aspects have meanwhile *forced* themselves on our attention, how remarkable is the change of attitude as evidenced by the finding of the late Royal Commission! [1896.]

From all this do not understand me to deride Science—for I have no intention of doing that; on the contrary, I think the debt we owe to modern investigation quite incalculable; but I only wish to warn you how complex all these problems are, how impossible that notion of settling even one of them by a cut-and-dried intellectual formula.

But you will ask (for this is the second point I mentioned some little time back) *how* people's emotions and feelings come in to colour their scientific conclusions? And the answer is—very simply, namely, by directing their choice as to what aspects of the problem they will ignore and what aspects they will envisage; by determining their point of view, in fact. To return to that illustration of several portrait-painters painting the same face; just as each painter is led by his feeling, his sympathies, his general temperament, to select certain points in the face and to pass over others, so each group of scientific men in each generation is led by its sympathies, its idiosyncrasies, to envisage certain aspects of the problems of the day and to ignore others.

The whole history of Science illustrates this. We are all familiar with the way in which the predilections of religious feeling in the time of Copernicus and Galileo retarded the progress of astronomical Science. As long as people believed that a divine drama of redemption had been enacted on this earth alone, they naturally concluded that this earth was the centre of the universe, and refused to look at facts which contradicted their conclusion. When Galileo turned his newly-made telescope on Jupiter and saw it circled by its satellites, he saw in this an image of the Copernican system and of the planets circling round the central Sun; but when he asked others to share his observation and his inference, they would not. "O, my dear Kepler," he writes in a letter to his fellow astronomer, "how I wish we could have one hearty laugh together. Here at Padua is the principal Professor of Philosophy, whom I have repeatedly and urgently requested to look at the moon and planets through my glass; but he pertinaciously refuses to do so. What shouts of laughter we should have at this glorious folly!"

And though we laugh at the folly of those before us, we do the same things ourselves to-day. Take the science of Political Economy. A revolution has taken place in that, almost comparable to the change from the geocentric to the heliocentric view in Astronomy. During the distinctively commercial period of the last 100 years, the leading students of social science, being themselves filled with the spirit of the time, have been fain to look upon the acquisition of private wealth as the one absorbing motive of human nature; and so it has come about that the economists, from Adam Smith to John Stuart Mill, have founded their science on self-seeking and competition, as the base of their analysis. To-day another series of economists coming to the front—their minds pre-occupied with the great facts of Community of life and Co-operation—have discovered that Society is in the main an illustration of these latter principles, and have evolved a quite new phase of the science. It is not that Society has changed so much during this period, as that the altered point of view of the students of Society has caused them simply to fix their attention on a different aspect of the problem and a different range of facts.

I have alluded already to the way in which the prevalent use of Machinery in practical life has affected our mental outlook on the world. It is curious that during this mechanical age of the last 100 years or so, we have not only come to regard Society in a mechanical light, as a concourse of separate individuals bound together by a mere cash-nexus, but have extended the same idea to the universe at large, which we look upon as a concourse of separate atoms, associated together by gravitation, or possibly by mere mutual impact. Yet it is certain that both these views are false, since the individuals who compose Society are *not* separate from each other; and the theory that the universe, in its ultimate analysis, is composed of a vast number of discrete atoms is simply unthinkable.

When we come to a practical and modern question like Medicine, the influence of the spirit in which it is approached on the course of the science is very easy to see. For if the science of Medicine is approached (as it perhaps mostly is to-day) in a spirit of combined Fear and Self-indulgence—fear for one's own personal safety, combined with a kind of anxiety to continue living in the indulgence of habits known to be unhealthy—if it is approached in this uncomfortable and contradictory state of mind, it is pretty obvious that its course will be similarly uncomfortable; that it will consist for the most part in a search for drugs which shall, without effort on our part, palliate the effects of our misconduct; in the discovery, as in a kind of nightmare, that the air round us is full of billions of microbes; in a terrified study of these messengers of disease, and in a frantic effort to ward them off by inoculations, vaccinations, vivisections, and so forth, without end.

If, on the other hand, the science is approached from quite a different side—from that of the love of Health, and the desire to make life lovely, beautiful and pure; if the student is filled not only with this, but with a great belief in the essential *power* of Man, and his command in creation, to control not only all these little microbes whose name is Legion, but through his mind all the processes of his body; then it is obvious enough that a whole series of different facts will arise before his eyes and become the subject of his study—facts of sanitation, of the laws of cleanly life, diet, clothing and so forth, methods of control, and the details and practice of the influence of the mental upon the physical part of man—facts quite equally real with the others, equally important, equally numerous perhaps and complex, but forming a totally different range of science.

In conclusion, you begin to see doubtless that I do not believe in a science of mere Formulas, which can be poured from one brain to another like water in a pot. I believe in something more organic to Humanity—which shall combine Sense, Intellect and Soul; which shall include the keenest training of the Senses, the exactest use of the Brain, and the subordination of both of these to the finest and most generous attitude of Man towards Nature.

To come to quite practical aspects, I think that Physical Science, and for that matter Natural History too, ought to be founded on the closest observation and actual intimacy with

Nature. It is notorious that in many respects the perceptions, the Nature-intuitions, of savage races far outdo those of civilised man. We have let that side go slack, and too often the man of science when he comes out of his study is a mere baby in the external world. I look back with a kind of shame when I think that I studied the mathematical side of Astronomy for three or four years at Cambridge and absolutely at the time hardly knew one star from another in the sky. But such are the methods of teaching that have been in use. They ought however to be reversed, and practical acquaintance with the facts should come a long way first, and then be succeeded by inductive and deductive reasoning when the difficulties of the subject have forced themselves on the student's mind.

Then in Natural History and Botany I think that we have hitherto not only neglected the perceptive side, but also what may be called the intuitive and emotional aspects. If any one will attend to the subject, I believe they will perceive that there are dormant in the mind the finest intuitions and instincts of relationship to the various animals and plants—intuitions which have played a far more important part in the life of barbaric races than they do to-day.[1] Primitive peoples have a remarkable instinct of the medicinal and dietetic uses of herbs and plants—an instinct which we also find well developed among animals—and I believe that this kind of knowledge would grow largely if, so to speak, it were given a chance. The formal classification of animals and plants—which now forms the main part of these sciences—would then come in simply as an aid and an auxiliary to the more direct and human study.

[1] Elisée Reclus, in his remarkable paper, *La Grande Famille*, points out the wide-reaching *Friendship*, and free alliance for various purposes, of primitive man with the animals, existing long before the so-called "domestication" of the latter. See *Humane Review*, Jan., 1906.

Again, let us take the science of Physiology. At present this is mainly carried on by means of Dissection or Vivisection. But both these methods are unsatisfactory. Dissection, because it amounts to studying the organisation of a living creature by the examination of its dead carcase; and Vivisection, because it is not only open to a similar objection, but because it necessarily violates the highest relation of man to the animal he is studying. There is, I believe, another method—a method which has been known in the East for centuries, though little regarded in the West—which may perhaps be called the method of Health. It consists in rendering the body, by proper habits of life, pure and healthy, till it becomes, as it were, transparent to the inner eye, and then projecting the consciousness *inward* so as to become almost as sensible of the structure and function of the various internal organs, as it usually is of the outer surface of the body. Of course this is a process which cannot be effectuated at once, and which may need help and corroboration by external methods of study, but I believe it is one which will lead to considerable results. There is no doubt that many of the Yogis of India attain to great skill in it.

Similarly, from what we have already said about Political Economy, it is obvious that satisfactory results in that science must depend immensely on the high degree of social instinct and feeling with which the student approaches it, and on the thoroughness of his acquaintance with the *actual life* of a people; and that the development of these factors is fully as important a part of the science as that which consists in the logical ordering and arrangement of the material obtained.

I need not, I think, go any further into detail of new methods in each Science. You remember what I said at the beginning about the Cell studying the Body of which it formed a part. We may imagine, if we like, three stages in this process. In the first stage the Cell regards the other cells and the Body simply from the point of view of how they affect *it*, and its comfort and safety. This might be taken to correspond to the Old-time Science. In the second stage the Cell, with its tiny experience of the other cells and the small part of the body in which it is placed, becomes highly intellectual, and professes to lay down the laws of the structure of the body generally. This corresponds to the attitude of Modern Science. In the third stage the Cell, growing and evolving, and coming daily into closer sympathetic relationship with all parts of the body, begins to find its true relation to the other cells, not to use *them*, but to fulfil its part in the whole. Gradually drawing all the threads together and coming more and more, so to say, into a central position, it at last in its little brain spontaneously and inevitably reflects the whole, and becomes the mirror of it. This would answer to what we have called a really rational and humane Science.

Man has to find and to *feel* his true relation to other creatures and to the whole of which he is a part, and has to use his brain to further this. Science *is*, as we all know, the search for Unity. That is its ideal. It unites innumerable phenomena under one law; and then it unites many laws under one higher; always seeking for the ultimate complete integration. But (is it not obvious?) Man cannot find that unity *of* the Whole until he feels his unity *with* the Whole. To found a Science of one-ness on the murderous Warfare and insane Competition of men with each other, and on the Slaughter and Vivisection of animals—the search for unity on the practice of disunity—is an absurdity, which can only in the long run reveal itself as such.

I do not know whether it seems obvious to you, but it does to me, that Man will never find in theory the unity of outer

Nature till he reaches in practice the unity of his own. When he has learnt to harmonise in himself all his powers, bodily and mental, his desires, faculties, needs, and bring them into perfect co-operation—when he has found the true hierarchy of himself—then somehow I think that Nature round him will reflect this order, and range itself in clear and intelligible harmony about him.

But I can say no more. I have dragged you by the neck, as it were, through a recondite and difficult subject; and even so I do not feel that I have by any means done justice to it. But it is possible, perhaps, that I have cast the germ of an idea among you, which, if you think over it at leisure, may develop into something of value.

THE END.

1278149

Made in the USA